DATA PROCESSING
MATHEMATICS

DATA PROCESSING

MATHEMATICS

George A. Gleim

Bucks County Community College
Newtown, Pennsylvania

Mario V. Farina

Telecommunications and Information Processing Operations
General Electric Company
Schenectady, N.Y.

PRENTICE-HALL, INC., Englewood Cliffs, New Jersey

© 1972 by
PRENTICE-HALL, INC.
Englewood Cliffs, New Jersey

10   9   8   7   6   5   4   3   2   1

ISBN: 0-13-196899-8
Library of Congress Catalog Card No. 70-37662

Printed in the United States of America

PRENTICE-HALL INTERNATIONAL, INC., London
PRENTICE-HALL OF AUSTRALIA, PTY. LTD., Sydney
PRENTICE-HALL OF CANADA, LTD., Toronto
PRENTICE-HALL OF INDIA PRIVATE LIMITED, New Delhi
PRENTICE-HALL OF JAPAN, INC., Tokyo

To Our Families

CONTENTS

Chapter 11

Chapter 12

Chapter 13

PREFACE

In developing the topics for a textbook covering the field of data processing mathematics, two major considerations appeared to be paramount. First, and foremost, was the issue concerned with what is relevant DP mathematics. The field of data processing is extremely plastic. It has only been a matter of a few years since emphasis in DP curriculums was focused on punched card technology. Now, however, programs are emphasizing third generation EDP hardware and software methods and capabilities. Whereas, in addition to training terminal-oriented students to program and operate DP machines, curriculums had been updated to provide tracts for those students who are interested in transferring to four-year schools where they can complete baccalaureate degrees in data processing, business administration, or other related programs. The emergence to prominence of courses in systems analysis made it appropriate to introduce such topics as blocking factors, decision tables, linear programming, and forecasting, in addition to other relevant DP topics including flowcharting, elementary programming, subscripting, numbers systems, and sorting concepts.

The second consideration was that of developing the topics in such a way that prior mathematical prerequisites were unnecessary. Therefore, while it would be desirable for the student using this text to have previously had a course in college algebra, the material in Chapter 1 should be sufficient to provide an adequate background in algebra. Notice that, while linear programming and forecasting could be developed using higher mathematics, only elementary algebra has been introduced while still providing the basic fundamental concepts involved in such new and dynamic topics.

The material in this text could be effectively covered in a three credit DP Math course, or perhaps integrated into introductory programming and systems analysis courses.

The authors are indebted to many individuals for assistance in preparing this book. To Mrs. Elsie Gleim and Mrs. Beatrice Shaffer go our thanks for typing the manuscript; to Karl Karlstrom goes the credit for suggesting this text; and to R.J. Hawke and W.B. Wright our gratitude for helping make the book possible.

Chapter 1

OVERVIEW OF ALGEBRA

Equations

Mathematical expressions, such as

GROSS PAY = HOURS × RATE

INTEREST = PRINCIPAL × RATE

FICA = GROSS PAY × .054

ON-HAND BALANCE = ON-HAND BALANCE + RECEIPTS – ISSUES

$T = N(L_I + 1)ms$

play a significant role in data processing. Most applications that the data processor will become involved with will in some way require a knowledge of algebra. Probably the most important concept in algebra is the idea of an equation.

An equation is an open sentence. If we replace the variables with elements that make the statement true, then we have a solution of the open sentence. Two equations are equivalent if they have the same solutions. The following properties are used in the manipulation of equations.

1. An equivalent equation is obtained if the same number is added to (or subtracted from) both members of an equation.

2. An equivalent equation is obtained if both members of an equation are multiplied (or divided) by the same number, provided that the number is not equal to zero.

For example, if

$$5X + 10 = 3X + 6$$

it is necessary only to transpose $+10$ and $3X$ by adding $-10 - 3X$ to both members of the equation. Thus

$$5X + 10 = 3X + 6$$
$$5X + 10 - 10 - 3X = 3X + 6 - 10 - 3X$$
$$2X = -4$$

Then dividing both sides by 2 we find the solution to be

$$X = -2$$

Note that division by zero is not permissible. Division by zero concludes in undesirable results. For example, consider the following algebraic exercise. Let

$$a = b$$

Therefore

$$ab = a^2$$
$$ab - b^2 = a^2 - b^2$$
$$b(a - b) = (a - b)(a + b)$$
$$b = a + b \qquad \text{(dividing by } a - b \text{, i.e., zero)}$$
$$b = 2b$$

Hence

$$1 = 2$$

EXERCISE

1-1. Solve the following equations:

(a) $5W = 4$

(b) $-3x = -6$

(c) $.2z = 1$

(d) $5 + R = 3R - 5$

(e) $\frac{2}{3}T = \frac{6}{9}$

(f) $\frac{3}{4}z + \frac{4}{7} = -\frac{3}{5} + \frac{1}{3}$

(g) $\frac{4}{17}x - \frac{3}{11} = \frac{6}{7}$

(h) $-.6R + 1.5 = -.5R - 2.5$

(i) $\dfrac{5}{x} = \dfrac{1}{3}$

(j) $\dfrac{5}{(x + 1)} = -7(-2x + 1)$

Linear Equations

An equation of the form

$$ax + by + c = 0$$

where not both a and b are zero, is a linear equation. Therefore

$$5x + y - 5 = 0$$
$$5x + 9 = 0$$
$$y - 8 = 0$$
$$y = 5x + 3$$

are linear equations. A linear equation can be put in the form

$$y = -\frac{a}{b}x - \frac{c}{b}$$

where

$$-\frac{a}{b}$$

is called the slope (direction) of the equation and

$$-\frac{c}{b}$$

is called the y-intercept of the equation.

Example 1. Given the equation

$$5x + y - 5 = 0$$

we can put the equation into the form

$$y = -5x + 5$$

to find the slope to be -5 and the y-intercept to be 5, as shown in Fig. 1-1.

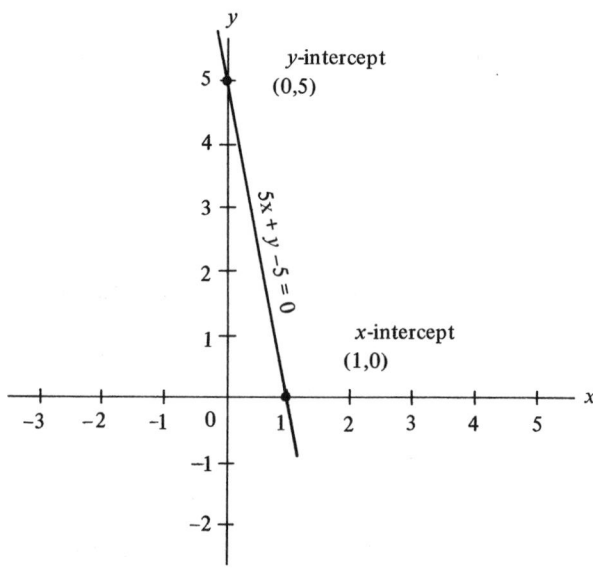

FIG. 1-1

The distance from the origin to the point where a linear equation crosses the x-axis is called the x-intercept of the equation. The x-intercept of the equation

$$5x + y - 5 = 0$$

is shown in Fig. 1-1 to be 1. A convenient method of drawing the graph of a linear equation is to find its x- and y-intercepts and to draw a straight line through the points where the axes are crossed.

Sometimes it is necessary to determine the equation of a straight line. If the slope and y-intercept are known, its equation can be readily written as

$$y = mx + b$$

where m is the slope and b is the y-intercept. This is known as the slope-intercept form of the equation of a straight line.

Another method for determining the equation involves utilizing the coordinates of two points on the line. Given two points on a straight line, $P_1:(x_1,y_1)$ and $P_2:(x_2,y_2)$, the equation of the line is

$$\frac{y - y_1}{x - x_1} = \frac{y_2 - y_1}{x_2 - x_1}$$

This method is known as the two-point form and is used later in the text to find equations of straight lines.

Example 2. Given the equation

$$2x - y + 1 = 0$$

we can put the equation into the form

$$y = 2x + 1$$

to find the slope to be 2 and the y-intercept to be 1, as shown in Fig. 1-2. Furthermore, the x-intercept can be readily found to be $-\frac{1}{2}$ by letting $y = 0$ and solving

$$0 = 2x + 1$$

Therefore we know two points on the line, namely $(-\frac{1}{2},0)$, and $(0,1)$. Using the two-point method we have

$$\frac{y - 0}{x + \frac{1}{2}} = \frac{1 - 0}{0 + \frac{1}{2}} = 2$$

and

$$y = 2x + 1$$

which is, of course, the equation we began with.

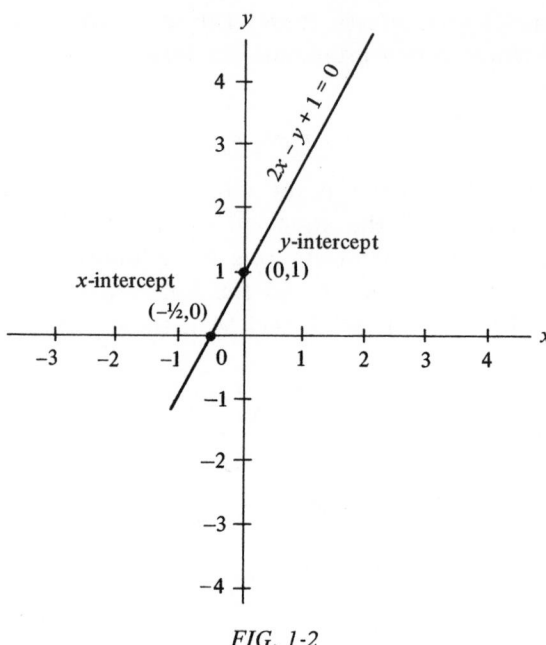

FIG. 1-2

EXERCISE

1-2. For each of the following linear equations find the slope, find the x- and y-intercepts, and draw the graph:

(a) $x + y - 1 = 0$

(b) $y = 2x + 3$

(c) $4x = y - 2$

(d) $2y = x$

(e) $x = 5$

(f) $-4y + 6 = 0$

(g) $-2x - y + 3 = 0$

(h) $\frac{2}{3}x - \frac{1}{2}y = 1$

(i) $y = 3x - 3$

(j) $2x - y = 6$

Solution of Two Simultaneous Linear Equations

Given two linear equations

$$ax + by = c$$
$$dx + ey = f$$

We have several methods available for finding the simultaneous solution.

To find the solution of two linear equations *graphically* we merely draw the straight lines that represent their equations and determine the coordinates of the point of intersection. If the lines coincide or are parallel, there is no one solution. The problem with this method is that the degree of accuracy is somewhat questionable.

Example 3. The lines representing linear equations

$$2x - y = 1$$
$$x + y = 2$$

are shown in Fig. 1-3. The point of intersection would appear to be approximately (1,1).

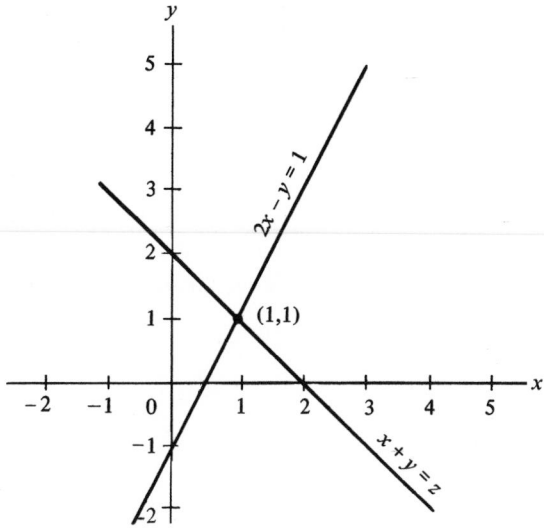

FIG. 1-3

A second method is elimination by substitution. This method requires that one of the unknowns is eliminated, resulting in a single equation in one unknown. For example, given the equations of Example 3,

$$2x - y = 1$$

$$x + y = 2$$

let us solve for y in terms of x in the first equation, getting

$$y = 2x - 1$$

Then substituting this value of y into the second equation we have

$$x + (2x - 1) = 2$$

$$3x - 1 = 2$$

$$3x = 3$$

and

$$x = 1$$

Substituting

$$x = 1$$

into any of the original equations allows y to be found. Hence

$$1 + y = 2$$

and

$$y = 1$$

A third method is elimination by addition or subtraction. Again, as in the previous method, we are attempting to eliminate one of the unknowns, leaving us with a single equation in one unknown. Given

$$2x - y = 1$$

$$x + y = 2$$

we can add the two equations, giving us an equation in x alone. Hence

$$3x = 3$$

and

$$x = 1$$

from which we can readily solve for y.

Finally, we may solve simultaneous equations by determinants. If $a, b, c,$ and d represent any numbers, the symbol

$$\begin{vmatrix} a & b \\ c & d \end{vmatrix}$$

is called a determinant of second order whose value is defined as

$$\begin{vmatrix} a & b \\ c & d \end{vmatrix} = ad - cb$$

Given two linear equations

$$ax + by = c$$

$$dx + ey = f$$

we can show, using the method of substitution, the solution to be

$$x = \frac{ce - fb}{ae - db}$$

$$y = \frac{af - dc}{ae - db}$$

Using the notation of determinants, therefore, we can also show the solution to be

$$x = \frac{\begin{vmatrix} c & b \\ f & e \end{vmatrix}}{\begin{vmatrix} a & b \\ d & e \end{vmatrix}}$$

$$y = \frac{\begin{vmatrix} a & c \\ d & f \end{vmatrix}}{\begin{vmatrix} a & b \\ d & e \end{vmatrix}}, \quad \text{if} \begin{vmatrix} a & b \\ d & e \end{vmatrix} \neq 0$$

Again, considering the equations of Example 3,

$$2x - y = 1$$

$$x + y = 2$$

and using the method of determinants we see that

$$x = \frac{\begin{vmatrix} 1 & -1 \\ 2 & 1 \end{vmatrix}}{\begin{vmatrix} 2 & -1 \\ 1 & 1 \end{vmatrix}} = \frac{1+2}{2+1} = 1$$

and

$$y = \frac{\begin{vmatrix} 2 & 1 \\ 1 & 2 \end{vmatrix}}{\begin{vmatrix} 2 & 1 \\ 1 & 1 \end{vmatrix}} = \frac{4-1}{2+1} = 1$$

EXERCISE

1-3. Solve each of the following sets of equations using four methods:

(a) $x + y = 4$
$x - y = -4$

(b) $x + y = 4$
$x + 2y = -8$

(c) $x - y = 6$
$2x - 3y = -8$

(d) $2x + 4y = -9$
$-3x + 5y = 1$

(e) $4x - 7y = 13$
$-3x + 2y = -5$

(f) $-x + 2y = 0$
$2x - 4y = -4$

(g) $y = 6$
$x - 2y = 4$

(h) $\dfrac{x}{2} + y = \dfrac{1}{2}$

$\quad 2x - \dfrac{y}{2} = 4$

(i) $R + S = 1.5$

$\quad -2.6R + 5.2S = -10.4$

(j) $ax + 2by = 4c$

$\quad 2ax - 3by = -6c$

Quadratic Equations

An equation of the form

$$ax^2 + bx + c = 0$$

is a quadratic equation. As with linear equations, there are various methods by which to arrive at solutions. First, it is possible to graph the function

$$y = ax^2 + bx + c = 0$$

to determine if the function crosses the x-axis at any point. If the curve crosses the x-axis at a point, the value of y at that point is zero, and the value of x at that point is a solution of the equation. If the function does not touch the x-axis, the equation has no real roots; if the function is tangent to the x-axis but does not cross it, the quadratic equation has only one root.

An algebraic method for solving a quadratic equation is by completing the square. That is, given a general quadratic equation

$$ax^2 + bx + c = 0$$

we use the following procedure to complete the square and find the solutions. First, transpose c:

$$ax^2 + bx + c = 0$$

$$ax^2 + bx = -c$$

Second, reduce the coefficient of x^2 to 1:

$$\frac{ax^2}{a} + \frac{bx}{a} = \frac{c}{a}$$

$$x^2 + \frac{b}{a}x = -\frac{c}{a}$$

Third, add half of the coefficient of x, squared, to both members of the equation:

$$x^2 + \frac{b}{a}x + \left(\frac{b}{2a}\right)^2 = -\frac{c}{a} + \left(\frac{b}{2a}\right)^2$$

$$x^2 + \frac{b}{a}x + \frac{b^2}{4a^2} = -\frac{c}{a} + \frac{b^2}{4a^2}$$

$$= -\frac{c}{a}\frac{4a}{4a} + \frac{b^2}{4a^2}$$

$$= -\frac{4ac}{4a^2} + \frac{b^2}{4a^2}$$

$$= \frac{b^2 - 4ac}{4a^2}$$

The expression on the left-hand side of the equation is a perfect square:

$$x^2 + \frac{b}{a}x + \frac{b^2}{4a^2} = \frac{b^2 - 4ac}{4a^2}$$

$$\left(x + \frac{b}{2a}\right)^2 = \frac{b^2 - 4ac}{4a^2}$$

Taking the square root of both sides we find

$$x + \frac{b}{2a} = \pm\frac{\sqrt{b^2 - 4ac}}{2a}$$

$$x = -\frac{b}{2a} \pm \frac{\sqrt{b^2 - 4ac}}{2a}$$

and

$$x = \frac{-b \pm \sqrt{b^2 - 4ac}}{2a}$$

which is known as the quadratic formula. Now, of course, we have also developed a third way of finding the solutions to a quadratic equation. That is, it is necessary only to substitute values of the coefficients a, b, and c into the formula.

Example 4. Solve the quadratic equation

$$2x^2 - x - 1 = 0$$

using multiple techniques.

Method 1

Let us draw the graph of the function

$$y = 2x^2 - x - 1$$

The graph of the function, Fig. 1-4, crosses the x-axis at the points $(1,0)$ and $(-\frac{1}{2},0)$. At these points the value of y is zero, and the values of x are solutions to the equation

$$2x^2 - x - 1 = 0$$

Verify by substituting $-\frac{1}{2}$ and 1 into the equation.

Method 2

Let us complete the square to find the solutions. Hence

$$2x^2 - x - 1 = 0$$
$$2x^2 - x = 1$$
$$x^2 - \tfrac{1}{2}x = \tfrac{1}{2}$$
$$x^2 - \tfrac{1}{2}x + \tfrac{1}{16} = \tfrac{1}{2} + \tfrac{1}{16}$$
$$(x - \tfrac{1}{4})^2 = \tfrac{9}{16}$$
$$= (\tfrac{3}{4})^2$$
$$x - \tfrac{1}{4} = \pm\tfrac{3}{4}$$
$$x = \tfrac{1}{4} \pm \tfrac{3}{4}$$

and

$$x = 1, -\tfrac{1}{2}$$

Method 3

Finally, substituting the values of the coefficients, $2, -1$, and -1 into the quadratic equation we have

$$x = \frac{1 \pm \sqrt{1 - 4(2)(-1)}}{4}$$

$$= \frac{1 \pm \sqrt{9}}{4}$$

$$= \frac{1 \pm 3}{4}$$

and

$$x = 1, \ -\frac{1}{2}$$

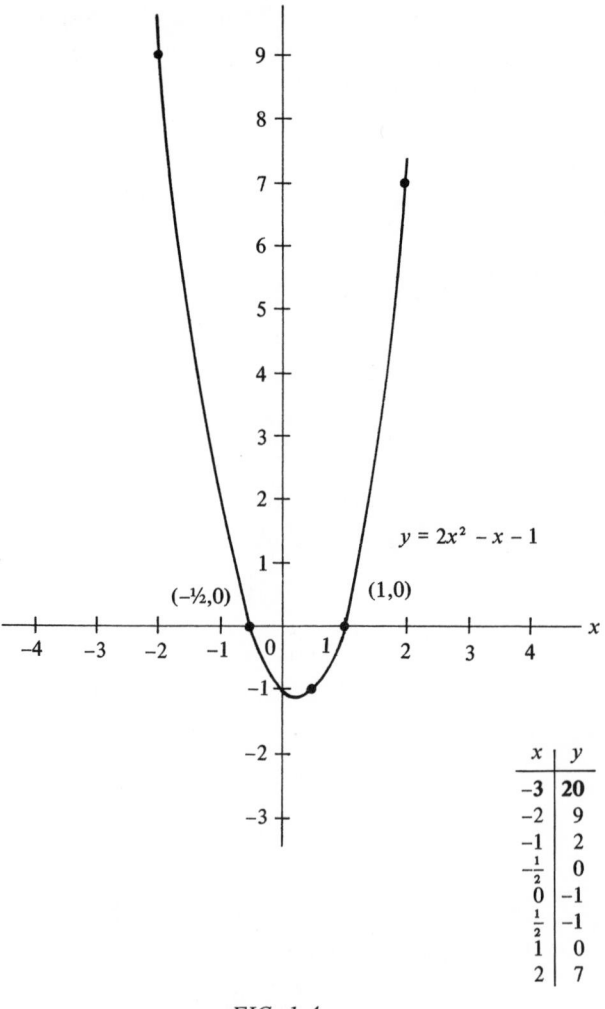

FIG. 1-4

The expression

$$b^2 - 4ac$$

under the radical of the quadratic equation is known as the discriminate. If the discriminate is negative, there are no real solutions. The solutions, in that case, are known as imaginary and will not be covered in this discussion.

EXERCISE

1-4. Solve the following quadratic questions using three methods:

(a) $x^2 + 2x - 3 = 0$

(b) $2y^2 - 5y - 12 = 0$

(c) $2y^2 + 3y = 2$

(d) $t^2 + 3t = 18$

(e) $4x^2 + 4x = 3$

(f) $4z^2 - 4 = 0$

(g) $2x^2 - 2x = 0$

(h) $x^2 - 1.5x = 3.0$

(i) $2.5x^2 + 1.0x = 5.0$

(j) $4.0x^2 = 9.0$

Progressions

A sequence of numbers is a group of numbers arranged in a definite order. Therefore

$$1, 2, 3, \ldots$$

$$1, 2, 4, 8, \ldots$$

$$2, 5, 8, \ldots$$

$$3, 6, 8, \ldots$$

represent sequences. An arithmetic *progression* is a sequence of numbers each of which, after the first, may be found by adding to the preceding number of the sequence the common difference between consecutive numbers of the sequence. Therefore, given the sequence

$$2, 5, 8, \ldots$$

we may find the fourth number of the sequence by adding the common difference, 3, to 8, or 11.

If a is the first term and d is the common difference of an arithmetic progression, the formula for the last term, L, is

$$L = a + (n - 1)d$$

where n is the number of elements in the sequence. The formula for the sum of the first n terms of an arithmetic progression is

$$S = \frac{n}{2}(a + L)$$

A geometric progression is a sequence of numbers each of which, after the first, may be found by multiplying the preceding number of the sequence by the common ratio. Therefore, the sequence

$$2, 4, 8, 16, \ldots$$

is a geometric progression in which the common ratio is 2. Hence the fifth element of the progression, 32, is found by multiplying the fourth term, 16, by 2.

The numbers of a geometric progression are

$$a, ar, ar^2, ar^3, \ldots$$

The formula for the last term is

$$L = ar^{n-1}$$

and the sum of the first n terms is

$$S = \frac{a(1 - r^n)}{1 - r}, \qquad \text{if } r \neq 1$$

Example 5. Given the arithmetic progression

$$1, 2, 3, \ldots$$

consisting of ten terms, find the last number, and, finally, find the sum of the terms of the progression. Therefore

$$L = 1 + (10 - 1)1$$
$$= 1 + 9$$

and

$$L = 10$$

Also,

$$S = \frac{10}{2}(1 + 10)$$
$$= 5(11)$$

and

$$S = 55$$

Example 6. Given the geometric progression

$$1, 2, 4, 8, \ldots$$

consisting of 8 terms, find the last number, and, finally, find the sum of the terms of the progression. Therefore

$$L = 1(2)^7$$

and

$$L = 128$$

Also,

$$S = \frac{1(1 - 2^8)}{1 - 2}$$

and

$$S = 255$$

EXERCISE

1-5. Find the specified term and sum of the following progressions:

    (a) Tenth term of 3, 6, 9, . . .

    (b) Tenth term of 3, 6, 12, . . .

    (c) Eighth term of 4, 7, 10, . . .

  (d) Eighth term of 4, 8, 16, . . .

  (e) Twentieth term of 2, 6, 10, . . .

  (f) Thirty-first term of 1, 2, 4, 8, . . .

  (g) Ninth term of $1, \frac{1}{2}, \frac{1}{4}, \ldots$

  (h) Twentieth term of $-1, -3, -5, \ldots$

  (i) Tenth term of $-1, 2, -4, 8, \ldots$

  (j) Thirty-first term of $-1, -1, -1, \ldots$

Applications

Example 7. This example demonstrates some of the calculations required in the processing of a weekly payroll register. Let us assume that 40 hours per week is standard with any excess of 40 hours considered to be overtime and calculated at one and one-half times the rate. The Federal Income Tax (FIT) is calculated at 14 percent of the taxable pay. Taxable pay is the difference between gross pay and nontaxable pay. Nontaxable pay is a function of both the number of dependents and tax exemption allowance. The social security (FICA) is calculated at 5.4 percent up to $8200 of gross income. Given the following information

Current hours	= 40.00
Rate	= $10.00
YTD earnings	= 0.00
YTD FIT	= 0.00
YTD FICA	= 0.00
No. dependents	= 4
Hospitalization	= $6.25
Pension	= $8.50
Credit union	= $5.25

we compute the net pay to be

Standard pay   = 40.00 × $10.00 = $400.00

Overtime pay   = 0.00

Gross pay   = $400.00

Nontaxable pay $= \dfrac{4 \times 650}{52} = \$50.00$

Taxable pay   = $400.00 – $50.00 = $350.00

FIT   = $350.00 × .14 = $49.00

FICA   = $400.00 × .054 = $21.60

Misc. deductions = $20.00

Net pay   = $400.00 – $49.00 – $21.60 – $20.00 = $309.40

Example 8. Compute the net pay as in Example 7. The difference in this problem is in the calculation of FICA. FICA is calculated at 5.4 percent of the gross pay up to $8200.00 of earnings. Therefore, given that

Current hours	= 80.00
Rate	= $10.00
YTD earnings	= $7500.00
YTD FIT	= $800.00
YTD FICA	= $405.00
No. dependents	= 12
Misc. deductions	= 0.00

we compute the net pay to be

Standard pay   = 40.00 × $10.00 = $400.00

Overtime pay   = 40.00 × $15.00 = $600.00

Gross pay   = $1000.00

Nontaxable pay	$= \dfrac{12 \times 650}{52} = \$150.00$
Taxable pay	$= \$1000.00 - \$150.00 = \$850.00$
FIT	$= .14 \times \$850.00 = \$119.00$
FICA	$= (\$8200.00 - \$7500).054 = \$37.80$
Net pay	$= \$1000.00 - \$119.00 - \$37.80 = \$843.20$

EXERCISE

1-6. Compute the net pay based on the following data:

Current hours	$= 75.00$
Rate	$= \$8.90$
YTD earnings	$= \$7,200.00$
YTD FIT	$= \$780.00$
YTD FICA	$= \$388.80$
No. dependents	$= 6$
Misc. deductions	$= \$27.75$

Example 9. A student has made scores of 35, 60, 65, 70, 75, and 80, respectively, in six examinations. These scores are shown in Fig. 1-5. The average score is found by

$$\text{AVERAGE} = \frac{\Sigma \text{ Scores}}{\text{Number of exams}}$$

$$= \frac{S_1 + S_2 + S_3 + S_4 + S_5 + S_6}{6}$$

$$= \frac{35 + 60 + 65 + 70 + 75 + 80}{6}$$

and

$$\text{AVERAGE} = 64.17$$

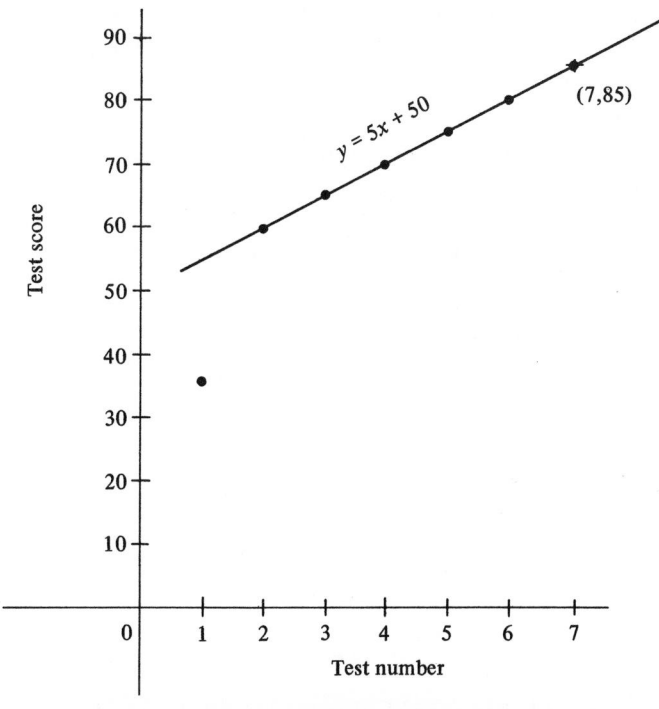

FIG. 1-5

The last five scores fall along a straight line. Using the two-point formula to determine the equation of the line we find

$$\frac{y - 60}{x - 2} = \frac{65 - 60}{3 - 2} = 5$$

$$y - 60 = 5(x - 2)$$

$$y - 60 = 5x - 10$$

and

$$y = 5x + 50$$

is a linear-equation passing through the points representing the last five examination scores. Therefore, if we felt it was appropriate to ignore the first test score, we might predict test score 7 to be 85. The concept of forecasting is given a more thorough discussion in Chapter

10. What we are trying to demonstrate here is the relevance of mathematical techniques that allow us to pass equations through points on a cartesian plane.

Example 10. Suppose that the volume of sales for a particular item was represented by the following historical data:

Month	Volume
January	10
February	15
March	20
April	25
May	30
June	35
July	40
August	45
September	50
October	fut.
November	fut.
December	fut.

We may ask the following questions: What will the predicted volume be for the last quarter? What will the total volume be for the entire year? And, can we develop an equation to represent the historical data?

The graph of Fig. 1-6 represents the data for the first three quarters of the year. Again, using the two-point formula we have

$$\frac{y - 20}{x - 3} = \frac{25 - 20}{4 - 3} = 5$$

$$y - 20 = 5(x - 3)$$

$$y - 20 = 5x - 15$$

and

$$y = 5x + 5$$

Therefore we might predict the sales for October, November, and December to be 55, 60, and 65 by substituting $x = 10$, 11, and 12, respectively, into the equation.

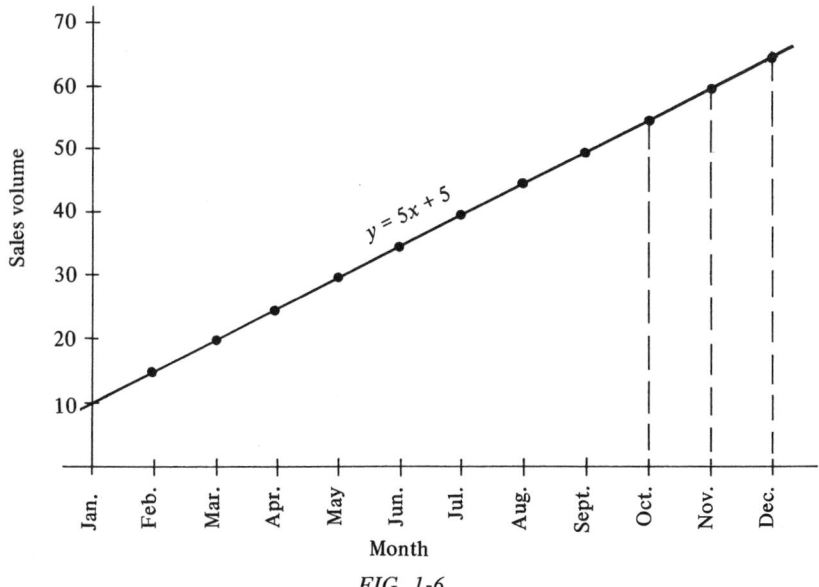

FIG. 1-6

Now consider the sequence formed by the volume of sales for the twelve-month interval:

10, 15, 20, 25, 30, 35, 40, 45, 50, 55, 60, 65

The sales clearly form an arithmetic progression. The sum of the terms of the progression represents the desired volume for the year. Therefore

$$\text{VOLUME} = \frac{12}{2}(10 + 65)$$

and

$$\text{VOLUME} = 390$$

The bar graph shown in Fig. 1-7 can also be used to represent the data of the example. The bar graph is introduced, at this point, because it plays a role in the chapter on forecasting, Chapter 10.

Example 11. The points representing the sales history in Example 10 were chosen so that they coincided with a linear equation. Generally, the data will not fall into this type of pattern, and some interpreting is necessary. Consider the following historical data:

Month	Volume
January	10
February	13
March	25
April	25
May	40
June	15
July	45
August	55
September	60
October	fut.
November	fut.
December	fut.

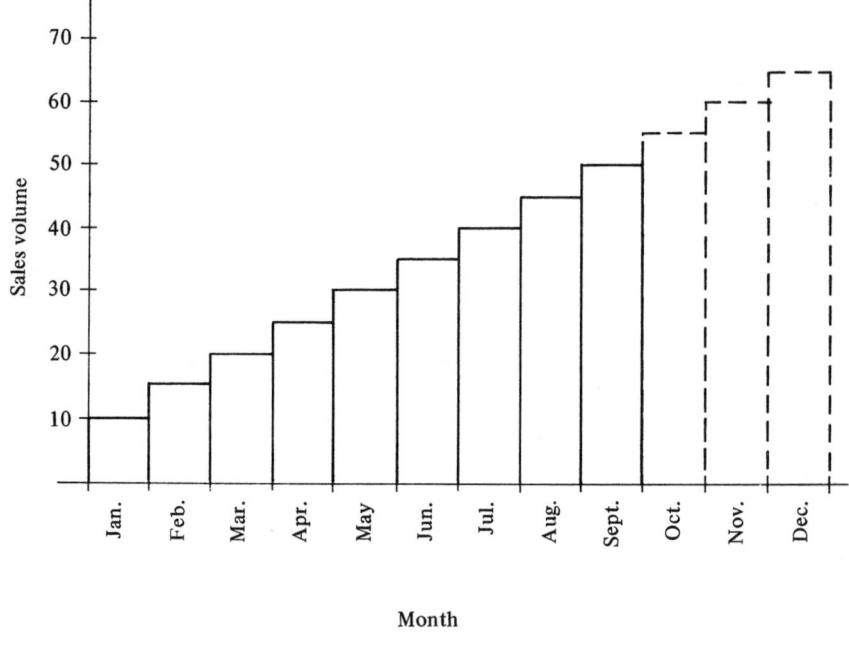

Month

FIG. 1-7

Figure 1-8 represents a linear equation that attempts to pass through the points in the best possible manner. It is only an approximation. The volume for June is assumed to be meaningless

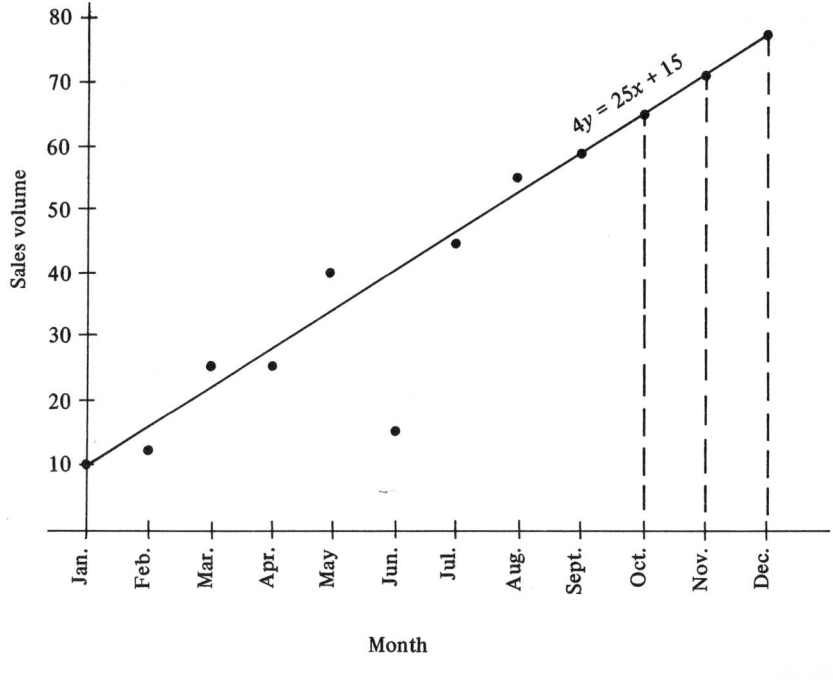

Month

FIG. 1-8

and is disregarded. The two-point formula uses the months of January and September. Hence

$$\frac{y - 10}{x - 1} = \frac{60 - 10}{9 - 1} = \frac{50}{8}$$

$$8(y - 10) = 50(x - 1)$$

$$8y - 80 = 50x - 50$$

$$4y - 40 = 25x - 25$$

and

$$4y = 25x + 15$$

Using this equation we are now able to approximate the sales for the remaining three months. Note that after getting the actual sales statistics for October, or October and November, a new equation could be developed for future forecasting.

Example 12. A programmer earns \$10,000 per year. If he can count on a 7 percent increase in salary for each of the next 5 years, what will his salary be and how much will he have earned after 5 years?

After 1 year	$SALARY_1$	$= \quad 10,000 + 10,000(.07)$
		$= \quad 10,000(1 + .07)$
	$SALARY_1$	$= \quad 10,000(1.07)$
After 2 years	$SALARY_2$	$= \quad 10,000(1.07) + 10,000(1.07)(1.07)$
		$= \quad 10,000(1.07 \, [1 + .07]$
		$= \quad 10,000(1.07)(1.07)$
	$SALARY_2$	$= \quad 10,000(1.07)^2$
After 3 years	$SALARY_3$	$= \quad 10,000(1.07)^2 + 10,000(1.07)^2(.07)$
		$= \quad 10,000(1.07)^2 \, [1 + .07]$
		$= \quad 10,000(1.07)^2 (1.07)$
	$SALARY_3$	$= \quad 10,000(1.07)^3$

$$\vdots \qquad\qquad\qquad \vdots$$

After 5 years	$SALARY_5$	$= \quad 10,000(1.07)^5$

$$\vdots \qquad\qquad\qquad \vdots$$

After n years	$SALARY_n$	$= \quad 10,000(1.07)^n$
GROSS EARNINGS		$= \quad SALARY_1 + SALARY_2 + \ldots + SALARY_5$
		$= 10,000(1.07) + 10,000(1.07)^2 + \ldots + 10,000(1.07)^5$

This sequence forms a geometric progression. The first term of the progression is $10,000(1.07)$, and the common ratio is 1.07. Therefore, using the formula to determine the sum of the terms of a geometric progression we have

$$S_{sum} = \frac{10,000(1.07)(1 - 1.07^5)}{1 - 1.07}$$

and

$$S_{sum} = \frac{10,000(1.07)(1.07^5 - 1)}{.07}$$

The actual computation of the programmer's salary after 5 years ($salary_5$) and the total money earned in 5 years (S_{sum}) is left as an exercise for the student.

EXERCISES

1-7. An automobile salesman has sold 10, 13, 16, 19, 22, 25, and 28 units in the first 7 months. How would you go about determining the sales for the remainder of the year? Total sales for the year?

1-8. Compute the following:
(a) $\text{Salary}_5 = 10,000(1.07)^5$

(b) $S_{\text{sum}} = \dfrac{10,000(1.07)(1.07^5 - 1)}{.07}$

1-9. Isaac came to America in 1575 and deposited $10 in the First National Bank of Chicago. His interest rate was 3 percent. He decided to move his family back to the old country in 1970 and withdrew his money from the bank. How much did he have?

1-10. A little boy had a wealthy uncle whom he considered to be a miser. On the boy's birthday the uncle gave the boy a choice of gifts. The boy could have a crisp new ten-dollar bill or a sum of pennies calculated in the following manner: one penny for the first day of the month, two pennies for the second day, four pennies for the third, eight pennies for the fourth, and so on for the 31 days in the month. Which gift should the boy choose?

Chapter 2

NUMBERS

Decimal System

We use numbers every day but perhaps have never really stopped to consider exactly what they mean. Consider the number 674 as an example. This number tells *how many* there are of certain things, be they dollars, cars, oranges, or pupils.

The number tells that there are six hundred and seventy-four of those things. Another way to think of 674 is as

6 one-hundreds

plus 7 tens

plus 4 units.

Proof:

6 × 100 = 600

7 × 10 = 70

4 × 1 = 4

Sum = 674

You can see that a number is made up of digits having positional values. In the number system that we use daily at the bank, in the supermarket, or at the office, the positional values can be illustrated by the chart in Fig. 2-1. Notice that, as you move to the left, each position is worth 10 times the value to its right. Now, given a certain

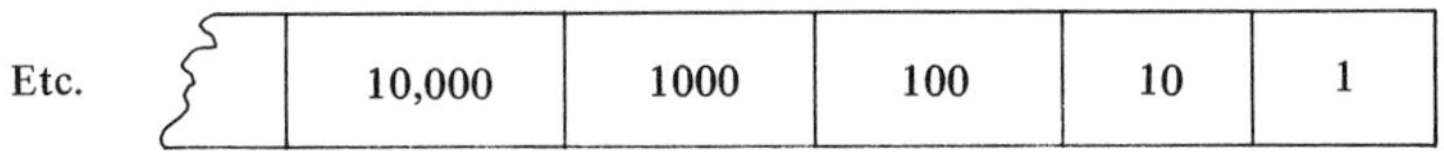

	10,000	1000	100	10	1

Etc.

FIG. 2-1

number of things, you can state how many there are by selecting appropriate divisions of the chart. Suppose that you have two-thousand three-hundred and nine dollars in the bank. Your bankbook shows

2309

If you place these digits above the rightmost divisions of the chart, Fig. 2-2 results.

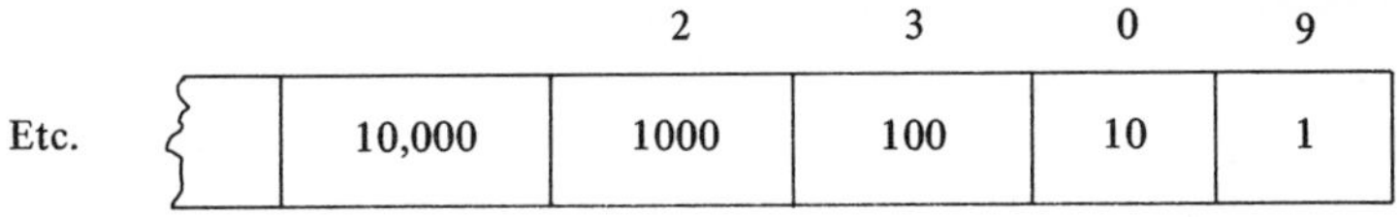

		2	3	0	9
	10,000	1000	100	10	1

Etc.

FIG. 2-2

If you wish, you may place leading zeroes ahead of 2309, but this is not necessary since people who work with numbers have agreed that if no digits are shown ahead of a number, those digits are assumed to be zero.

Now, can you see that 2309 means

2 one-thousands

plus 3 one-hundreds

plus 0 tens

plus 9 units

Proof:

$$2 \times 1000 = 2000$$
$$3 \times 100 = 300$$
$$0 \times 10 = 0$$
$$9 \times 1 = 9$$

Sum = 2309

The number system we have been discussing is called the *decimal system*. The word *decimal* is derived from the Latin word meaning "ten." The decimal system employs ten digits. They are 0, 1, 2, 3, 4, 5, 6, 7, 8, and 9. We may express values as large or as small as we please using decimal numbers. Thus 6385 could be considered a relatively large number, while -7938 could be considered a relatively small number.

Why do we use ten digits in our number system? Probably because human beings have ten fingers.

Base 6 System

Numbers that are based upon ten digits are often called *base 10* numbers. But base 10 is not the only base that may be employed when you wish to tell how many there are of various things. If you wish, you may employ numbers based upon two digits (*base 2*), three digits (*base 3*), four digits (*base 4*), etc.

Imagine a fictitious country called Maniland where all the inhabitants have only three fingers on each hand. Instead of employing the base 10 system that we use, they may prefer a system based upon six digits, those digits being 0, 1, 2, 3, 4, and 5. When the citizens of Maniland write numbers they use positional values obtained by the chart in Fig. 2-3. Excepting the rightmost position, each position is worth 6 times the value of the position to its right.

Etc.	7776	1296	216	36	6	1

FIG. 2-3

How many things would a person in Maniland be talking about when he writes the number

35044

Placing the digits of the number above the rightmost divisions of the chart, Fig. 2-4 results. The chart indicates that the Maniland

		3	5	0	4	4
Etc.	7776	1296	216	36	6	1

FIG. 2-4

inhabitant is indicating the value made up of the sum of three 1296s, five 216s, zero 36s, four 6s, and four 1s. Or,

$3 \times 1296 = 3888$

$5 \times 216 = 1080$

$0 \times 36 = 0$

$4 \times 6 = 24$

$4 \times 1 = 4$

Sum $= 4996$

When discussing numbers to various bases, it is important to specify *which base* we mean so that confusion is kept to a minimum. It is customary to indicate a base by writing a *subscript* to the right of a number. Thus 35044_6 means that the number is written in the base 6 system, and 4996_{10} means that the number is written in the base 10 system. From the previous example, we may say that

$$35044_6 = 4996_{10}$$

From this point, we shall clearly indicate in which base various numbers are written by appending subscripts to those numbers. *If a number does not have a subscript, you may assume that it is written in the base 10 system.*

Let us convert another number from the base 6 system to the base 10 (decimal) system: 310045_6. Place the number above the base

3	1	0	0	4	5
Etc. } 7776	1296	216	36	6	1

FIG. 2-5

6 chart (Fig. 2-5). The numbers shown within the divisions of the chart are decimal values. To obtain the decimal equivalent of 310045_6, we make these computations:

$$3 \times 7776 = 23328$$
$$1 \times 1296 = 1296$$
$$0 \times 216 = 0$$
$$0 \times 36 = 0$$
$$4 \times 6 = 24$$
$$5 \times 1 = 5$$

$$\text{Sum} = 24653_{10} \qquad (310045_6 = 24653_{10})$$

EXERCISE

2-1. Give the base 10 equivalents of these base 6 numbers:

(a) 324_6

(b) 50025_6

(c) 3_6

(d) 4000_6

(e) 11001_6

(f) 230550_6

(g) 30003_6

(h) 555_6

(i) 12345_6

(j) 1355421_6

Base 5 System

Let us try a problem where the number to be converted to the base 10 system is

$$3412_5$$

This is a base 5 number and is converted using a base 5 chart (Fig. 2-6). As before, the numbers shown within the divisions of the chart are base 10 values. Observe that as you move toward the left, the values in the divisions are multiplied by 5.

		3	4	1	2
Etc.	625	125	25	5	1

FIG. 2-6

The computations to be made are these:

$3 \times 125 = 375$

$4 \times 25 = 100$

$1 \times 5 = 5$

$2 \times 1 = 2$

$\text{Sum} = 482_{10} \qquad (3412_5 = 482_{10})$

EXERCISE

2-2. Convert these base 5 numbers to their base 10 equivalents:
  (a) 100_5

  (b) 43_5

  (c) 3_5

  (d) 40004_5

  (e) 444_5

(f) 1234_5

(g) 3410032_5

Constructing Base Charts

When you need to convert from any base to base 10, you must construct a chart that shows the decimal values for each of the divisions. The rightmost digit of the chart is always 1 (Fig. 2-7). The

				1

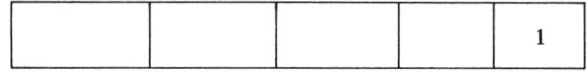

FIG. 2-7

next digit to the left is the *same as the base being converted*. Suppose that the base is 8 (Fig. 2-8). The value placed in this division is the multiplier to use for all remaining divisions to the left (Fig. 2-9).

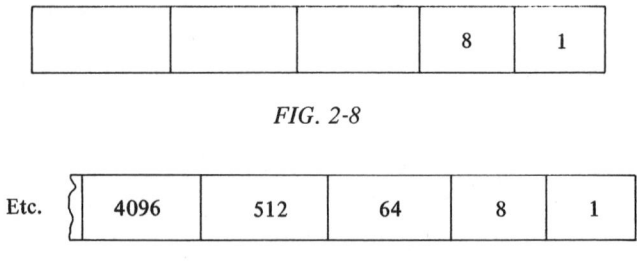

			8	1

FIG. 2-8

Etc.	4096	512	64	8	1

FIG. 2-9

A more rigorous way to show the values that must be shown in the divisions is as in Fig. 2-10. If b, the base, is 16, the values to place within the divisions are $16^0 = 1$, $16^1 = 16$, $16^2 = 256$, $16^3 = 4096$, etc.

Etc.	b^3	b^2	b^1	b^0

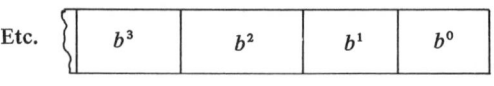

FIG. 2-10

Base 16 System

It is certainly possible to express numbers in bases greater than 10: bases 12, 16, and 60 are examples. When doing so you are faced with the problem of inventing symbols to stand for values greater than 9. Consider base 16, which is a base useful in data processing. This base requires 16 symbols. Ten familiar ones are available from our experience: 0, 1, 2, 3, 4, 5, 6, 7, 8, and 9. We need six more. We could use "exotic" symbols such as +, &, !, #, <, and > but in practice, we find that there is a minimum of confusion if we use the letters A, B, C, D, E, and F.

Now we can prepare a table containing the digits used in base 16 and their decimal equivalents:

Base 16 Digits

Symbol Used in Base 16	Decimal Equivalents
0	0
1	1
2	2
3	3
4	4
5	5
6	6
7	7
8	8
9	9
A	10
B	11
C	12
D	13
E	14
F	15

And we are able to express numbers in base 16 like this:

$$4B3F5_{16}$$

To determine the decimal equivalent of this number, let us

		4	B	3	F	5
Etc.	1,048,576	65536	4096	256	16	1

FIG. 2-11

prepare a base 16 chart and place the digits of the number above it (Fig. 2-11). Now we make the usual computations, keeping in mind that B in the number stands for decimal 11 and F stands for decimal 15:

$$4 \times 65536 = 262144$$
$$11 \times 4096 = 45056$$
$$3 \times 256 = 768$$
$$15 \times 16 = 240$$
$$5 \times 1 = 5$$

$$\text{Sum} = 308213_{10} \qquad (4B3F5_{16} = 308213_{10})$$

EXERCISE

2-3. Give the base 10 equivalents of these base 16 numbers:

(a) 8_{16}

(b) $34B_{16}$

(c) ABC_{16}

(d) $400A_{16}$

(e) 8000_{16}

(f) FFF_{16}

(g) 1010_{16}

Converting a Base 10 Number to Base 8

There are times when we must convert from a number expressed in decimal to a number expressed in some other base. Suppose, for example, that we need to convert 478_{10} to a base 8 number. A method we can use is to divide 478_{10} repeatedly by 8. Study these calculations:

Remainders give answer

```
            0        7
    8    | 7         3
    8    | 59        6
    8    | 478₁₀
```

We divide 478 by 8. The result is 59 with a remainder of 6. Note where the remainder is placed. Then we divide 59 by 8. The result is 7 with a remainder of 3. The 3 is written above the 6. Finally, we divide 7 by 8. This gives a result of zero with a remainder of 7. The 7 is placed above the 3.

The answer to the problem is now available. It is 736_8. Therefore

$$478_{10} = 736_8$$

If we wish, we may check the answer. Converting 736_8 to decimal proves that our calculations were correct (Fig. 2-12):

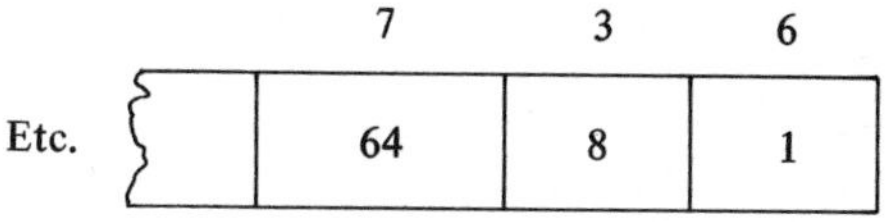

	7	3	6
Etc.	64	8	1

FIG. 2-12

$7 \times 64 = 448$

$3 \times 8 = 24$

$6 \times 1 = 6$

Sum $= 478_{10}$ $(736_8 = 478_{10})$

EXERCISE

2-4. Convert these base 10 numbers to base 8:
   (a) 99_{10}

   (b) 123_{10}

(c) 1001_{10}

(d) 4_{10}

(e) 9_{10}

(f) 2000_{10}

(g) 8_{10}

Converting a Base 10 Number to Base 16

Here is an example where 2754_{10} must be converted to base 16. The constant divisor in our calculations is 16.

```
                                ┌──────── Remainders give answer
                    0         ↓10
      16  ┌── 10         12
      16  ┌── 172         2
      16  ┌── 2754₁₀
```

Dividing 2754_{10} by 16 gives the result 172 with a remainder of 2; dividing 172 by 16 gives the result 10 with a remainder of 12; and dividing 10 by 16 gives a result of zero with a remainder of 10. The answer to the problem is, therefore, $AC2_{16}$. (Recall that in base 16, A stands for decimal 10 and C stands for decimal 12.)

Let us check the results (see Fig. 2-13):

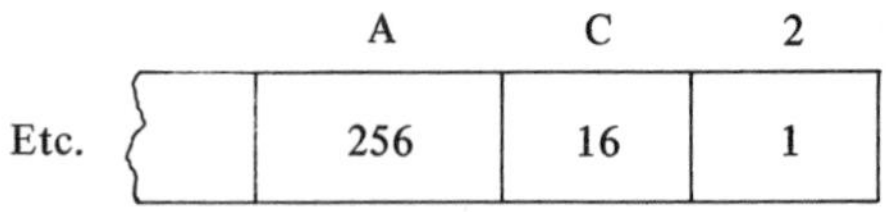

	A	C	2
Etc. {	256	16	1

FIG. 2-13

$$10(A) \times 256 = 2560$$

$$12(C) \times 16 = 192$$

$$2 \times 1 = 2$$

$$\text{Sum} = 2754_{10} \qquad (AC2_{16} = 2754_{10})$$

EXERCISE

2-5. Convert these base 10 numbers to base 16:

(a) 16_{10}

(b) 10_{10}

(c) 24_{10}

(d) 394_{10}

(e) 99_{10}

(f) 2000_{10}

(g) 1101_{10}

Converting a Decimal Number to Any Base

When you want to convert *any* decimal number to any other base, a method you can use, which always works, is to divide the decimal number by a number representing the desired base. That is, if you want to convert a number from base 10 to base 6, you divide repeatedly by 6; if you want to convert a number from base 10 to base 12, you divide repeatedly by 12; etc.

Can you see *why* this method works? Consider the decimal number 5683_{10}. What does the number really tell us? As you have already seen, the number represents a sum of numbers. The sum consists of

$$5 \text{ groups of } 1000\text{s} = 5000$$

$$\text{plus } 6 \text{ groups of } 100\text{s} = 600$$

$$\text{plus } 8 \text{ groups of } 10\text{s} = 80$$

$$\text{plus } 3 \text{ groups of } 1\text{s} = \underline{3}$$

$$\text{Sum} = 5683_{10}$$

Should we need to convert 5683_{10} to base 7, our calculations must show how many groups of 16807s are needed, how many groups of 2401s are needed, etc. The values for the groups mentioned are obtained from the base 7 chart shown in Fig. 2-14.

We can now see that no groups of 16807s are needed; therefore

| Etc. | 16807 | 2401 | 343 | 49 | 7 | 1 |

FIG. 2-14

we place zero above the division labeled 16807. Next we determine that two groups of 2401 are needed; therefore we place 2 above the division labeled 2401. So far, we have the result shown in Fig. 2-15.

| | 0 | 2 | | | | |
| Etc. | 16807 | 2401 | 343 | 49 | 7 | 1 |

FIG. 2-15

Now, twice 2401 is 4802. Subtracting 4802 from 5683, we find that we must select the proper multiples of 343, 49, 7, and 1 to make up 881. Can you see why it would have been wrong to place 3 above the division labeled 2401? Answer: because 2401 X 3 is 7203, a number larger than 5683. Could we have gotten away with merely placing 1 above the division labeled 2401? No, because that would have left 5683 – 2401 = 3282 to account for using values 343, 49, 7, and 1. It is impossible to select base 7 digits zero through 6 to place above the divisions labeled with those values and still make up 3282. It would not be correct to place digits 7 or larger above those divisions because base 7 numbers use only the digits zero through 6.

When we correctly place 2 above the division labeled 2401, we must make up the difference between 5683 and 4802, which we have seen is 881. How many 343s are needed? The answer is 2 because 343 X 2 = 686, while 343 X 3 = 1029 (too large). Subtracting 686 from 881 gives 195. How many 49s are needed? The answer is 3. Three times 49 is 147, and 147 subtracted from 195 gives 48. How many 7s are needed? The answer is 6. Six times 7 is 42 and 42 from 48 gives 6. The digit 6 is placed above the division labeled 1 (Fig. 2-16) and the problem is solved. Therefore $5683_{10} = 22366_7$. Let us check our work by converting 22433_7 to decimal. Using Fig. 2-16, we make these calculations:

| | 2 | 2 | 3 | 6 | 6 |
| Etc. | 16807 | 2401 | 343 | 49 | 7 | 1 |

FIG. 2-16

$2 \times 2401 = 4802$

$2 \times 343 = 686$

$3 \times 49 = 147$

$6 \times 7 = 42$

$6 \times 1 = 6$

Sum $= 5683_{10}$ $(5683_{10} = 22366_7)$

Note that in two instances the digits to place above the divisions in the chart were 6's. When using this method, you need never fear that digits higher than the base allows will be required. The reason is easy to understand if you will examine the base 7 chart. Is it possible that the rightmost division (the one labeled 1) will *ever* require a 7 to be placed above it? No, because if 7 is the remaining portion of a base 10 number to account for, the entire remaining portion would have to be claimed by the division labeled 7. In that case, the division labeled 1 would have received zero.

We indirectly use the method just described when we convert a base 10 number to its base 7 equivalent by dividing repeatedly by 7. Study the calculations needed to convert 5683_{10} to base 7:

                               Remainders give answer

            0   2

      7 ⌐ 2   2

      7 ⌐ 16   3

      7 ⌐ 115   6

      7 ⌐ 811   6

      7 ⌐ 5683_{10}

The calculations show that there are 811 7s with 6 1s left over. To determine how many 49s are represented by 811 7s, another division is made by 7. (Two divisions by 7 is, of course, equivalent to a single division by 49.) The result shows that 811 7s represent 115 49s with 6 7s left over. The remainders are the values that give the final answer. Thus $(6 \times 1) + (6 \times 7) + (3 \times 49) + (2 \times 343) + (2 \times 2401) + (0 \times 16807)$ gives the final answer.

EXERCISE

2-6. Convert

(a) 3504_6 to base 10

(b) 1001_8 to base 10

(c) $3B6_{16}$ to base 10

(d) 10201_3 to base 10

(e) 1011_2 to base 10

(f) 340_5 to base 10

(g) 1246_7 to base 10

(h) 777_8 to base 10

(i) 684_{12} to base 10

(j) 39_{20} to base 10

Converting a Number in Any Base to Its Equivalent in Any Other Base

It is possible to convert a number expressed in *any* base to another number expressed in *any other* base. There are two ways you can do this. The first way, which may seem to be the more involved, is actually the more straightforward. Convert the number to base 10, and then convert it to the new base.

An example: We would like to convert the number 546_7 to its equivalent in base 5. First convert 546_7 to base 10 by employing the base 7 chart (Fig. 2-17):

		5	4	6
Etc.	343	49	7	1

FIG. 2-17

$5 \times 49 = 245$

$4 \times 7 = 28$

$6 \times 1 = \underline{\quad 6}$

Sum $= 279_{10}$ $(546_7 = 279_{10})$

Now convert 279_{10} to base 5:

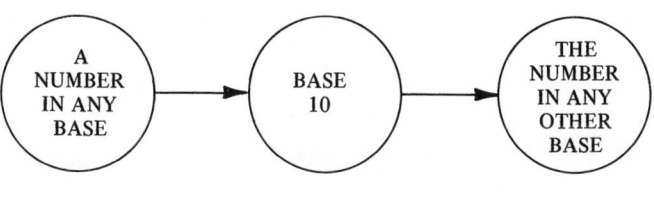

Remainders give answer

$279_{10} = 2104_5$; therefore $546_7 = 2104_5$.

Figure 2-18 shows a procedure that may always be employed to convert a number as it is expressed in any base to its equivalent in another base.

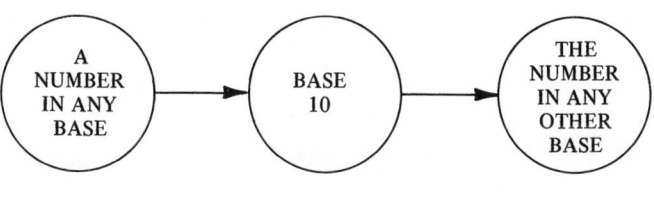

| A NUMBER IN ANY BASE | → | BASE 10 | → | THE NUMBER IN ANY OTHER BASE |

FIG. 2-18

EXERCISE

2-7. Convert
   (a) 33_6 to base 5
   (b) $AB6_{16}$ to base 8
   (c) $AB6_{16}$ to base 4
   (d) 101_{10} to base 3
   (e) 3777_8 to base 10
   (f) 456_8 to base 2
   (g) 824_{20} to base 16
   (h) 13_{12} to base 7
   (i) 11111_2 to base 16
   (j) 20002_3 to base 9

Converting Directly from Base to Base

It is possible to convert directly from base 7 to base 5. The base 7 number must be divided repeatedly by 5_7 and the work has to be done in base 7 arithmetic.

Let us take the same problem: convert 546_7 to base 5. Before we can attack the problem, we must prepare base 7 addition and multiplication tables. Figure 2-19 is the base 7 addition table. Use

	0	1	2	3	4	5	6
0	0	1	2	3	4	5	6
1	1	2	3	4	5	6	10
2	2	3	4	5	6	10	11
3	3	4	5	6	10	11	12
4	4	5	6	10	11	12	13
5	5	6	10	11	12	13	14
6	6	10	11	12	13	14	15

Note: All numbers shown in this table are in the Base 7 System.

FIG. 2-19 Base 7 addition table.

the base 7 addition table exactly the same way that you use the decimal (base 10) addition table. That is, find the first digit to add on an outside column, find the second digit on an outside row, and then look up the answer where the column and row intersect. We can see that $4_7 + 6_7$ is 13_7, that $5_7 + 2_7$ is 10_7, etc.

For practice, add this series of numbers in base 7:

$$3643_7$$
$$2165_7$$
$$1602_7$$
$$2653_7$$
$$2004_7$$
$$\overline{}$$

$$\text{Sum} = \quad ?$$

The sum obtained from the rightmost column is obtained by first adding $3_7 + 5_7$. This gives 11_7. Then add 11_7 and 2_7. This gives 13_7. Then add 13_7 and 3_7. This gives 16_7. Finally, add 16_7 and 4_7. The result is 23_7. Put down 3_7 and carry 2_7.

$$
\begin{array}{r}
2 \\
3643_7 \\
2165_7 \\
1602_7 \\
2653_7 \\
2004_7 \\
\hline
\end{array}
$$

$$\text{Sum} = \qquad 3_7$$

Sum the other columns using the same method. When you are all done, you should have the result 16033_7.

You may check your answer by converting all the numbers in the problem to base 10, summing them, and then checking the answer to see if it is the base 10 equivalent of 16033_7. The decimal equivalents of the five numbers in the series are 1354, 782, 639, 1018, and 690. The sum is 4483.

	0	1	2	3	4	5	6
0	0	0	0	0	0	0	0
1	0	1	2	3	4	5	6
2	0	2	4	6	11	13	15
3	0	3	6	12	15	21	24
4	0	4	11	15	22	26	33
5	0	5	13	21	26	34	42
6	0	6	15	24	33	42	51

Note: All numbers shown in this table are in the Base 7 System.

FIG. 2-20   Base 7 multiplication table.

Figure 2-20 is the base 7 multiplication table. The table shows that $4_7 \times 3_7$ is 15_7, that $5_7 \times 6_7$ is 42_7, etc.

Here is a base 7 multiplication problem:

$$346_7$$
$$\times 435_7$$

$$\text{product} = \quad ?$$

Your work should look like this:

$$346_7$$
$$\times \quad 435_7$$
$$2432_7$$
$$1404_7$$
$$2053_7$$

$$225102_7$$

Now convert 346_7 and 435_7 to base 10; multiply them and check the product against the base 10 equivalent of 225102_7. The decimal equivalents of the values being multiplied are 181 and 222. The product is 40182.

Let us go back to the problem that started this discussion about base 7 addition and multiplication tables. Convert 546_7 to base 5. We must divide repeatedly by 5_7 and do all the work in base 7 arithmetic. Here is the work:

```
                                  ┌──────── Remainders give answer
               0         2 ◄┘
        5  │ 2       1
        5  │ 14      0
        5  │ 106     4
        5₇ │ 546₇
```

Therefore $546_7 = 2104_5$. This answer agrees with the result that we obtained when 546_7 was converted to a base 10 intermediate result and the intermediate result was converted to base 5.

When dividing 546_7 by 5_7, you must use the base 7 multiplication table to determine the individual digits of the result. For instance, the table shows that 5_7 goes into 5_7 exactly once. That explains the 1 in the result (106 with a remainder of 4). Now, 5_7 will not go into 4_7 but it will go into 46_7. It will go into that number 6 times with a remainder of 4.

When dividing 106_7 by 5_7, you find that 5_7 goes into 10_7 once with a remainder of 2 (because $1_7 \times 5_7 = 5_7$ and $5_7 + 2_7 = 10_7$). Now, 5_7 goes into 26_7 exactly 4 times. The result is, therefore, 14_7 with a remainder of 0_7.

Dividing 14_7 by 5_7 and 2_7 by 5_7 should present no problems.

All numbers mentioned in the preceding three paragraphs were in the base 7 system.

Base 16

Base 16 is an important base where data processing is concerned. Let us try a few problems involving conversions of numbers from base 16 to various other bases.

		3	7	E
Etc.	4096	256	16	1

FIG. 2-21

Problem 1. Convert $37E_{16}$ to base 9 using base 10 as an intermediate base (see Fig. 2-21).

$$3 \times 256 = 768$$

$$7 \times 16 = 112$$

$$14(E) \times 1 = \underline{14}$$

$$\text{Sum} = 894_{10} \quad (37E_{16} = 894_{10})$$

Remainders give answer

$$
\begin{array}{c|c|c}
 & 0 & 1 \\
9 & 1 & 2 \\
9 & 11 & 0 \\
9 & 99 & 3 \\
9 & 894_{10} & \\
\end{array}
$$

$894_{10} = 1203_9$. Therefore $37E_{16} = 1203_9$.

Problem 2. This problem is the same Problem 1 except convert $37E_{16}$ *directly* to base 9. You will recall that when 546_7 was converted directly to base 5, all the work was done in base 7 arithmetic. To solve this problem, the work must be done in base 16 arithmetic. Use the base 16 multiplication table shown in Figure 2-22.

	0	1	2	3	4	5	6	7	8	9	A	B	C	D	E	F
0	0	0	0	0	0	0	0	0	0	0	0	0	0	0	0	0
1	0	1	2	3	4	5	6	7	8	9	A	B	C	D	E	F
2	0	2	4	6	8	A	C	E	10	12	14	16	18	1A	1C	1E
3	0	3	6	9	C	F	12	15	18	1B	1E	21	24	27	2A	2D
4	0	4	8	C	10	14	18	1C	20	24	28	2C	30	34	38	3C
5	0	5	A	F	14	19	1E	23	28	2D	32	37	3C	41	46	4B
6	0	6	C	12	18	1E	24	2A	30	36	3C	42	48	4E	54	5A
7	0	7	E	15	1C	23	2A	31	38	3F	46	4D	54	5B	62	69
8	0	8	10	18	20	28	30	38	40	48	50	58	60	68	70	78
9	0	9	12	1B	24	2D	36	3F	48	51	5A	63	6C	75	7E	87
A	0	A	14	1E	28	32	3C	46	50	5A	64	6E	78	82	8C	96
B	0	B	16	21	2C	37	42	4D	58	63	6E	79	84	8F	9A	A5
C	0	C	18	24	30	3C	48	54	60	6C	78	84	90	9C	A8	B4
D	0	D	1A	27	34	41	4E	5B	68	75	82	8F	9C	A9	B6	C3
E	0	E	ГC	2A	38	46	54	62	70	7E	8C	9A	A8	B6	C4	D2
F	0	F	1E	2D	3C	4B	5A	69	78	87	96	A5	B4	C3	D2	E1

Note: All numbers shown in this table are in the Base 16 System.

FIG. 2-22   Base 16 multiplication table.

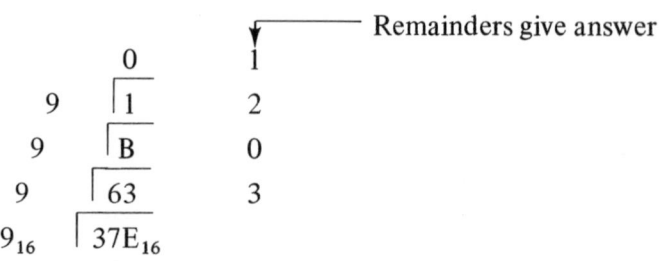

The answer is $37E_{16} = 1203_9$.

Problem 3. Convert $37E_{16}$ directly to base 8.

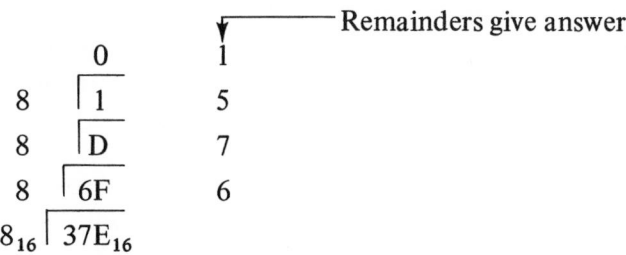

```
                        ┌──────── Remainders give answer
            0           1
   8    │ 1            5
   8    │ D            7
   8    │ 6F           6
  8₁₆   │ 37E₁₆
```

$$
\begin{array}{r|l}
 & 0 \qquad 1 \\
8 & 1 \qquad 5 \\
8 & D \qquad 7 \\
8 & 6F \qquad 6 \\
8_{16} & 37E_{16} \\
\end{array}
$$

Problem 4. Convert $37E_{16}$ directly to base 2.

$$
\begin{array}{r|l}
 & 0 \qquad 1 \\
2 & 1 \qquad 1 \\
2 & 3 \qquad 0 \\
2 & 6 \qquad 1 \\
2 & D \qquad 1 \\
2 & 1B \qquad 1 \\
2 & 37 \qquad 1 \\
2 & 6F \qquad 1 \\
2 & DF \qquad 1 \\
2 & 1BF \qquad 0 \\
2_{16} & 37E_{16} \\
\end{array}
$$

Remainders give answer

Later when we discuss more fully the special relationships between bases 2, 8, and 16, you will see that numbers may be converted from any of these bases to any other of these bases more simply than by the method shown above.

Base 2

Numbers expressed in base 2 are not only interesting but also important in the field of data processing. Virtually all computers built today represent numbers internally in base 2 (binary) form. This is true whether computer users refer to those numbers in their base 8, 10, or 16 forms.

In Chapter 3, where we discuss elementary computer concepts, we shall give additional details regarding why binary numbers are important to computers. In this chapter, we shall simply study their organization and learn how to work with them.

Binary numbers are numbers written in base 2. Hence there are only two digits in binary numbers, zero and 1. Despite the fact that only two digits are used in expressing a number in binary form, we are nevertheless able to express any value we please, providing we are willing to use enough zeroes and ones.

Consider this binary number:

$$10111011_2$$

What is its value? It is easy to find its value if we first prepare a binary chart and then write the binary number above it, one digit per division, as in Fig. 2-23.

		1	0	1	1	1	0	1	1
Etc.		256	128	64	32	16	8	4	2	1

FIG. 2-23

Now we make these calculations:

$$1 \times 128 = 128$$
$$0 \times 64 = 0$$
$$1 \times 32 = 32$$
$$1 \times 16 = 16$$
$$1 \times 8 = 8$$
$$0 \times 4 = 0$$
$$1 \times 2 = 2$$
$$1 \times 1 = \underline{1}$$

$$\text{Sum} = 187_{10} \quad (10111011_2 = 187_{10})$$

This procedure should be very familiar to you by now. We have used it in converting values from base 5 to decimal, base 7 to decimal, base 8 to decimal, etc. Now we employ it to convert base 2 to decimal.

You can see that the digits in a binary number have positional values. Those values, reading from right to left, are 1, 2, 4, 8, 16, etc., and correspond to powers of 2: 2^0, 2^1, 2^2, 2^3, 2^4, etc.

The binary chart may show either the actual values corresponding to the binary positions, or the powers of two form, or both (Fig. 2-24).

Etc.	2^7 128	2^6 64	2^5 32	2^4 16	2^3 8	2^2 4	2^1 2	2^0 1

FIG. 2-24

Sometimes you will be able to convert a binary number to its decimal equivalent almost at sight. Consider this number:

$$101000_2$$

You will see 101 in this number and recognize it as 5_{10}. Double 5, giving 10; double 10 (giving 20); double 20 (giving 40). You have now computed the decimal equivalent of 101000_2. The value is 40_{10}. Did you see how this was done? There were *three* doubles, one for each digit to the right of 101.

Let us try another example:

$$101100_2$$

You will recognize that 1011_2 represents 11_{10}. Double twice. (There are *two* digits to the right of 1011.) The result is 44_{10}.

Another example:

$$1100101_2$$

You will recognize 11_2, which represents 3_{10}. Double 5 times because there are five digits to the right of 11_2. The result is 96_{10}. This is not the complete answer because the problem is

$$1100101_2 \quad \text{not } 1100000_2$$

All that still must be done is to add to 96_{10} the decimal equivalent of 101_2 found at the right of 1100101_2. The value is 5_{10}. Now, $96_{10} + 5_{10}$ gives 101_{10}, which is the decimal equivalent of 1100101_2.

This technique of converting a binary value to its equivalent decimal value cannot always be used. But, given a problem, examine the number to see if the technique is applicable. The trick is to look for any familiar combination near the left-hand of the binary number and then add any value found at the right-hand end of the number. In this number,

$$11000011_2$$

the trick will work if you notice that 11_2 represents 3_{10} or that 110_2 represents 6_{10} or that 1100_2 represents 12_{10}, etc. If you elect to use 11_2, you will double 6 times; if you elect to use 110_2, you will double 5 times; if you elect to use 1100_2, you will double 4 times; etc. In any case, the final result of the doublings will be 192_{10}. To get the answer to the problem, you must add 3_{10} to 192_{10}, giving 195_{10}.

When dealing with binary numbers, you will have occasion to add, subtract, multiply, and even divide them. To do so, you must learn the binary addition and multiplication tables (Figs. 2-25 and 2-26). The tables are very simple and can be memorized in just a few minutes.

	0	1
0	0	1
1	1	10

Note: All numbers shown in this table are in the Base 2 System.

FIG. 2-25 Binary addition table.

	0	1
0	0	0
1	0	1

Note: All numbers shown in this table are in the Base 2 System.

FIG. 2-26 Binary multiplication table.

The tables show that

$0 + 0 = 0$

$0 + 1 = 1$

$1 + 0 = 1$

$1 + 1 = 10$

$0 \times 0 = 0$ *Note*: All numbers shown in this tabulation are in the base 2 system.

$0 \times 1 = 0$

$1 \times 0 = 0$

$1 \times 1 = 1$

There are no other possible single-digit combinations of values to add or multiply.

Let us try an addition problem. Add 11011_2 and 11010_2. Here is the work:

$$11011_2$$
$$11010_2$$
$$\overline{110101_2}$$

As you can see, a "carry" is necessary when 1_2 is added to 1_2. In one place, the example also shows that 1_2 must be added to 1_2 and that a 1_2 carry must also be added. Series of 1s may be easily handled if you take them two at a time. Therefore $1_2 + 1_2$ is 0_2 with a 1_2 carry. Adding 1_2 to that zero gives a final result for the column of 1_2.

Check your answer by converting 11011_2, 11010_2, and 110101_2 to decimal and checking the results.

Now a multiplication problem. Multiply 10011_2 by 1101_2:

$$10011_2$$
$$1101_2$$
$$\overline{10011}$$
$$00000$$
$$10011$$
$$10011$$
$$\overline{11110111_2}$$

Did you notice that except for the bottom row of numbers, the work contains the same digits as if it had been a base 10 multiplication problem?

Be sure to check the results in the usual way (by converting 10011_2, 1101_2, and 11110111_2 to decimal and checking the results).

Suppose that it is necessary for you to convert a decimal number to its binary equivalent. Take 483_{10}, for example. Simply divide 483_{10} by 2 repeatedly:

Remainders give answer

	0	1
2	1	1
2	3	1
2	7	1
2	15	0
2	30	0
2	60	0
2	120	1
2	241	1
2	483_{10}	

Therefore $483_{10} = 111100011_2$.

Base 2 Fractions

Everything we have discussed so far has had to do with integers (whole numbers). You may wish to express a value that has an integer part and a fractional part. Or you may wish to express a pure fraction. Let us consider the latter problem first.

What is the decimal equivalent of this next base 2 number:

$$.1011001_2$$

The position of the first digit to the right of the decimal point is worth 2^{-1} or .5; the second digit, 2^{-2} or .25; the third digit, 2^{-3} or .125; etc.

We may compute the value of the number as

$1 \times 2^{-1} = .5$
$0 \times 2^{-2} = 0$
$1 \times 2^{-3} = .125$
$1 \times 2^{-4} = .0625$
$0 \times 2^{-5} = 0$
$0 \times 2^{-6} = 0$
$1 \times 2^{-7} = \underline{.0078125}$

   Sum = $.6953125_{10}$ $(.1011001_2 = .6953125_{10})$

The base 2 chart may, therefore, be expanded to the right so that it looks as in Fig. 2-27.

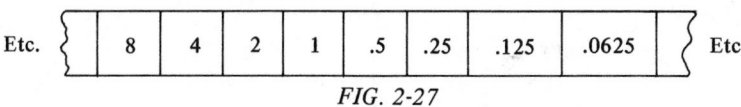

FIG. 2-27

Now, given any base 2 number, including those that contain fractions, you should be able to convert them easily to base 10. Example:

$$1011.1001_2$$

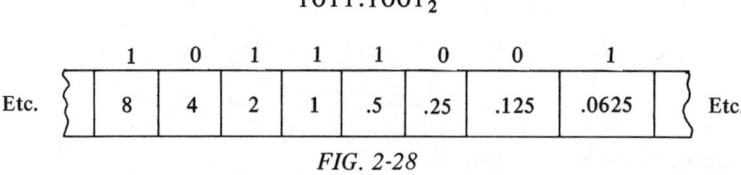

FIG. 2-28

Place the number over the base 2 to base 10 conversion chart (Fig. 2-28) and make these calculations:

$1 \times 8 \quad = 8$
$0 \times 4 \quad = 0$
$1 \times 2 \quad = 2$
$1 \times 1 \quad = 1$
$1 \times .5 \quad = .5$
$0 \times .25 \quad = 0$
$0 \times .125 \quad = 0$
$1 \times .0625 = \underline{.0625}$

   Sum = 11.5625_{10} $(1011.1001_2 = 11.5625_{10})$

Notice that using the binary chart, one places the binary number above the divisions in such a way that the decimal point is placed at the boundary line between the 2^1 and 2^{-1} divisions.

For more complicated numbers, the base 8 and base 16 number systems may, of course, be used to convert base 2 numbers to base 10. Let us use base 8 for this number:

$$1010100101.1110101_2$$

Break up the number into groups of three digits working right and left from the decimal point and write the equivalent base 8 digits (Fig. 2-29) obtained from this table:

Base 2 to Base 8 Conversion Chart

Binary Digits	Base 8 Digits
000	0
001	1
010	2
011	3
100	4
101	5
110	6
111	7

$$\underline{001}\ \underline{010}\ \underline{100}\ \underline{101}\ .\ \underline{111}\ \underline{010}\ \underline{100}_2$$
$$\ \ 1\ \ \ \ \ 2\ \ \ \ \ 4\ \ \ \ \ 5\ \ .\ \ 7\ \ \ \ 2\ \ \ \ \ 4_8$$

Now make these computations:

1 X 512	= 512	
2 X 64	= 128	
4 X 8	= 32	
5 X 1	= 5	
7 X .125	= .875	
2 X .015625	= .031250	
4 X .001953125 =	.0078125	

$$\text{Sum} = 677.9140625_{10}$$

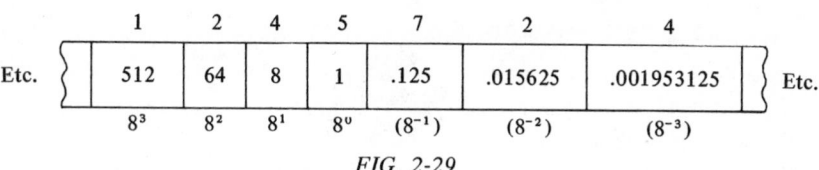

	1	2	4	5	7	2	4	
Etc. {	512	64	8	1	.125	.015625	.001953125	} Etc.
	8^3	8^2	8^1	8^0	(8^{-1})	(8^{-2})	(8^{-3})	

FIG. 2-29

$1245.724_8 = 677.9140625_{10}$; therefore $1010100101.1110101_2 = 677.9140625_{10}$

The same problem may be solved using the base 16 chart. First break up the number into groups of four digits and write the corresponding base 16 digits obtained from this table:

Base 2 to Base 16 Conversion Chart

Binary	Base 16 Digit
0000	0
0001	1
0010	2
0011	3
0100	4
0101	5
0110	6
0111	7
1000	8
1001	9
1010	A
1011	B
1100	C
1101	D
1110	E
1111	F

0010	1010	0101	.	1110	1010_2
2	A	5		E	A

	2	A	5 .	E	A	
Etc. {	256	16	1	.0625	.00390625	} Etc.
	16^2	16^1	16^0	16^{-1}	16^{-2}	

FIG. 2-30

Now place the number $2A5.EA_{16}$ over the base 16 to base 10 conversion chart (Fig. 2-30) and make these computations:

$$2 \quad \times 256 \qquad = 512$$

$$10(A) \times \ 16 \qquad = 160$$

$$5 \quad \times \ 1 \qquad = \ 5$$

$$14(E) \times .0625 \qquad = \qquad .8750$$

$$10(A) \times .00390625 = \qquad .0390625$$

$$\text{Sum} = 677.9140625_{10}$$

This result agrees with the one previously calculated.

We may convert a base 10 fraction to its binary equivalent directly by multiplying repeatedly by 2, or indirectly by multiplying repeatedly by 4, 8, 16, 32, etc. Below we shall give examples using multiples of 2, 8, and 16.

Suppose that we have the fraction $.833_{10}$, which must be converted to base 2. Here are the computations needed when the multiplier is 2_{10}:

$$
\begin{array}{c|l}
. & 833_{10} \\
& 2 \\
\hline
1 & 666 \\
& 2 \\
\hline
1 & 332 \\
& 2 \\
\hline
& 664 \\
& 2 \\
\hline
1 & 328 \\
& 2 \\
\hline
& 656 \\
& 2 \\
\hline
1 & 312 \\
\end{array}
\qquad (.833_{10} = .110101_2)
$$

Observe that a vertical line has been placed extending downward from the decimal point location in the value $.833_{10}$. Now, everything to the *right* of the line is multiplied by 2. The result is, of course, 1.666. The digits to the right of the decimal point are written at the right of the vertical line, while the 1 is written to the left of the line. We say that the 1 *spills over* the line.

Now repeat the process. Be careful to multiply only .666 by 2, not 1.666. When .666 is multiplied by 2, the result is 1.332. The digit 1 spills over to the left of the vertical line.

The next time the process is repeated, .332 (not 1.332) is multiplied by 2. This gives a result of .664. There is no spillover.

The solution to the problem may now be obtained by writing the spillover digits. These are $.110101_2$. Note that spillover *zeroes* are written where blanks are shown. Also note that this answer is not complete and precise. The value of $.833_{10}$ does not *exactly* equal 110101_2. This is true because we could have continued indefinitely multiplying by 2 if we had wished. In many problems *exact* binary equivalents to decimal fractions cannot be obtained, but you may make the conversion as close as you please by continuing to multiply by 2.

Now let us solve the same problem using the base 8 system as a stepping stone. Multiply repeatedly by 8_{10}:

Spillovers give ⟶⌐
      answer   ↓

$$
\begin{array}{c|l}
 . & 833_{10} \\
   & 8 \\
\hline
6 & 664 \\
  & 8 \\
\hline
5 & 312 \\
  & 8 \\
\hline
2 & 469
\end{array}
\qquad (.833_{10} = .652_8 = .110101010_2)
$$

Basically the process is identical to the one used when you multiply by 2. Spillover digits are in base 8, though. You first obtain the base 8 equivalent of the decimal fraction and then convert the octal digits to binary.

In the solution, we carried out the solution to more binary places than we did when we multiplied by 2.

Now, let us use the base 16 system:

Spillovers give
answer

$$
\begin{array}{r|l}
. & 833_{10} \\
 & 16 \\
\hline
4 & 998 \\
8 & 33 \\
\hline
13 & 328 \\
 & 16 \\
\hline
1 & 968 \\
3 & 28 \\
\hline
5 & 248 \\
 & 16 \\
\hline
1 & 488 \\
2 & 48 \\
\hline
3 & 968
\end{array}
$$

$(.833_{10} = D53_{16} = .110101010011_2)$

Observe carefully that when you multiply by 16_{10}, it is possible to have a two-digit spillover. Recall that in base 16 the decimal value 13 is represented by D_{16} and that D's binary equivalent is 1101_2. In this example the spillover digits are often two digits. To make it easier for you to determine which are the spillover digits, we have printed them in boldface type.

EXERCISE

2-8. Convert

(a) $.34_{10}$ to base 2

(b) $.84_{10}$ to base 8

(c) .05 to base 16

Converting Mixed Numbers

Now convert this number to binary:

$$474.326_{10}$$

First convert the integer portion and then the fractional portion. Let us use the base 8 system as the stepping stone:

$$\text{Remainders give answer}$$

$$\begin{array}{rcl} 0 & & 7 \\ 8 \,\lceil\, 7 & & 3 \\ 8 \,\lceil\, 59 & & 2 \\ 8_{10} \,\lceil\, 474_{10} & & (474_{10} = 732_8 = 110111010_2) \end{array}$$

Spillovers give answer

$$\begin{array}{r|l} . & 326 \\ & 8 \\ \hline 2 & 608 \\ & 8 \\ \hline 4 & 864 \\ & 8 \\ \hline 6 & 912 \\ & 8 \\ \hline 7 & 296 \end{array}$$

$(.326_{10} = .2467_8 = .010100110111_2)$

Therefore 474.326_{10} equals approximately 111011010.010100110111_2.

The procedures for converting a decimal number to binary may be used to convert a decimal number to any base. For example, suppose that we wish to convert $.1603_{10}$ to base 3. We do this by multiplying $.1603_{10}$ repeatedly by 3:

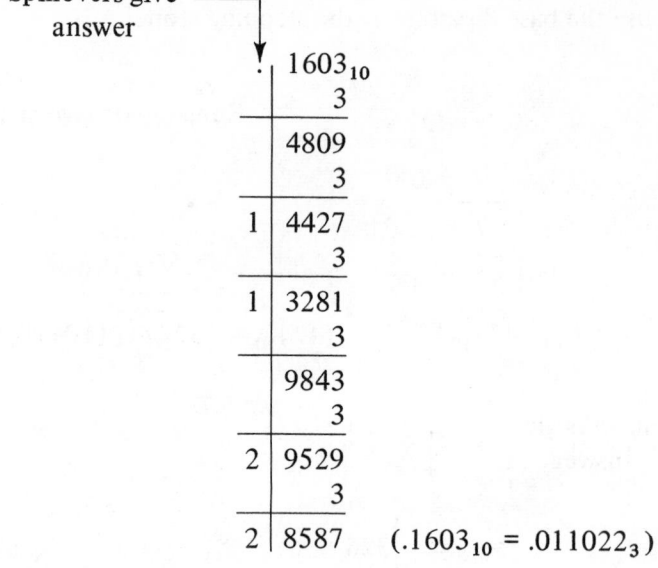

Spillovers give
answer

$.1603_{10}$
3
4809
3
1 | 4427
3
1 | 3281
3
9843
3
2 | 9529
3
2 | 8587 $(.1603_{10} = .011022_3)$

We may check the answer by preparing and using a base 3 fraction chart (Fig. 2-31):

$$0 \times 3^{-1} \ (.333333) = 0$$

$$1 \times 3^{-2} \ (.111111) = \ .111111$$

$$1 \times 3^{-3} \ (.037037) = \ .037037$$

$$0 \times 3^{-4} \ (.012345) = 0$$

$$2 \times 3^{-5} \ (.004115) = \ .008230$$

$$2 \times 3^{-6} \ (.001372) = \ \underline{.002744}$$

$$\text{Sum} = \ .159122_{10}$$

	0	1	1	0	2	2	
Etc.	3^{-1}	3^{-2}	3^{-3}	3^{-4}	3^{-5}	3^{-6}	Etc.

Fig. 2-31

$.1603_{10}$ is approximately equal to 011022_3. If we wish greater accuracy, we may obtain it by calculating to more places.

EXERCISES

2-9. How many one-thousands are there in the number 8478_{10}? How many one-hundreds, tens, units?

2-10. There are how many different digits employed in the decimal numbering system? In the octal? Binary? Base 16?

2-11. What are the decimal equivalents of the following base 6 numbers:
(a) 3452_6

(b) 204515_6

(c) 500040_6

2-12. What are the decimal equivalents of the following base 5 numbers:
(a) 40003_5

(b) 1234_5

(c) 4444_5

2-13. Construct a base 9 chart that may be used when converting base 9 numbers to decimal.

2-14. What are the decimal equivalents of the following base 16 numbers:
(a) 304_{16}

(b) ABC_{16}

(c) $4F5E_{16}$

2-15. Convert
(a) 3074_{10} to base 8

(b) 267_{10} to base 5

(c) $6EC4_{16}$ to base 7

2-16. Convert
(a) $2B38_{16}$ to base 2

(b) $2B38_{16}$ to base 4

(c) $2B38_{16}$ to base 8

2-17. Construct the base 5 addition table.

2-18. Construct the base 5 multiplication table.

2-19. Give the sum of these base 5 numbers:

$$1234$$
$$3004$$
$$2103$$
$$0323$$
$$\underline{4112}$$

2-20. Give the product of the base 5 numbers 344 and 214.

2-21. Give the decimal equivalents of the following binary numbers:
(a) 00110101

(b) 01100011

(c) 00011101

(d) 01111111

2-22. What is the binary sum of binary numbers 00111011 and 00100101? Of 00001111 and 01111111? Of 00101001 and 00111011?

2-23. What is the binary product of the binary numbers 010110 and 001110?

2-24. What is the decimal equivalent of the binary fraction .01001?

2-25. What is the decimal equivalent of the binary number 01001.00101?

2-26. What is the binary equivalent of the decimal fraction .328125?

2-27. What is the binary equivalent of the decimal value 81.53125?

2-28. What is the binary equivalent of the base 16 value 3B.A6?

Chapter 3

COMPUTER ARITHMETIC

A Clerk

A digital computer is a device that processes information. In this sense it is no more than a giant clerk. A clerk receives information, examines it, sorts it, makes calculations, writes summaries, etc. A computer does the same things.

If you would like to know whether a computer can do a particular job, ask yourself whether a clerk, given several lifetimes if necessary, could do the same task. If the answer is yes, then a computer could, most likely, do the job.

A computer is not a magical device. It can make millions of calculations in one second, but it cannot tell you what XYZ stock will do tomorrow. It can compute and print hundreds of paychecks in a single minute, but it cannot tell you whether you will receive an A for this course.

In this chapter you will learn how a computer stores numbers. Follow the discussion step by step, working out the exercises as you do so, and you should have little difficulty in mastering the material. As you go through the material keep in mind that you will only occasionally have to convert a decimal number to one of its binary forms. Sometimes the conversion task is tedius and it is well to remember that most conversions are made by the computer itself. Nevertheless, if you know *how* a computer stores numbers and converts from one form to another, you will be in a better position to understand what a computer is doing whenever it becomes necessary for you to debug a program. (Debug: find out why a program does not work and make corrections.)

Memory

A computer stores numbers in a place called its *memory*. A computer's memory is made up of units called locations. Each location may hold either a piece of data (such as a number) or a computer instruction. We shall talk about computer instructions later. For now, let us discuss numbers.

Figure 3-1 shows how you may visualize computer locations. How many memory locations a computer has and how large each location is depends on the computer you are studying. A typical computer may have in excess of 64,000 locations, each location having the capacity to hold a small number such as − 2345.6792 or a large number such as 394732.81. These examples are not maximums by any means. Maximums depend entirely on which computer is being discussed.

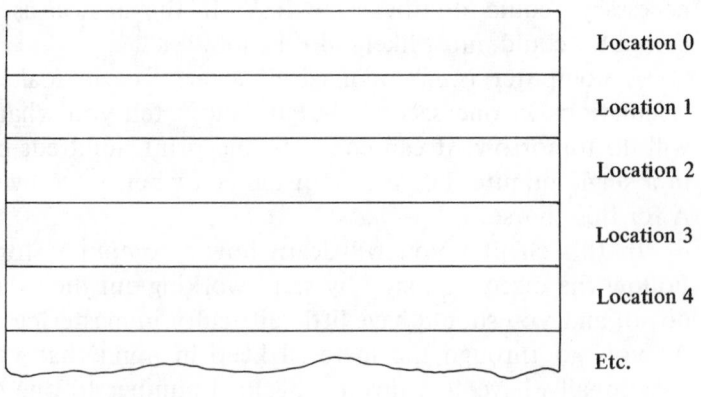

Etc.

FIG. 3-1

Numbers

Numbers stored in memory locations are stored in binary form (base 2). Again, the number of binary digits in a memory location varies with the computer you are dealing with. It is common to find computers that have 20, 24, 32, 36, 48, and 64 binary digits per location.

A binary digit is called a *bit*. Therefore, when we say a computer location has a capacity of 32 bits, we mean that it uses 32 zeroes and ones to express numbers. Thus 32 bits may be used to express the values 117, 14894, and 1972483 in this way:

$$00000000000000000000000001110101_2 \quad (117_{10})$$

$$00000000000000000011101000101110_2 \quad (14894_{10})$$

$$00000000000111100001100100000011_2 \quad (1972483_{10})$$

In many computers, the term *byte* means a group of 8 bits. Therefore, if a computer has a 65536-location memory and each location contains 32 bits, the capacity of the memory is 65536×4 or 262,144 bytes. (There are 4 bytes per 32-bit location.)

Often the number of locations in a computer's memory is a power of 2. Therefore you may hear about computers that have 4096, 8192, 16384 locations, etc.

Each location in a computer's memory has its own identifying serial number: 0, 1, 2, 3, 4, etc. This identifying number is called the location's *address*.

In some computers, *each byte* is individually addressed. This means that the memory is not organized by locations but by bytes. Each individual byte has its own address: 0, 1, 2, 3, 4, etc.

A computer that is organized by locations is called a *word* machine and the memory is said to have 32K words, 64K words, etc. (A K represents 1024 locations.) A computer that is organized by bytes is called a byte machine. Its capacity in bytes is also expressed as 16K, 32K, 64K, etc.

The figures given above are typical but there are exceptions. For instance, some computers are addressed by individual units called *characters*, not bytes, and each character may have only 4 or 6 bits. You should be aware of the fact that there is wide variation in the computer industry regarding the organization of memories.

There are two basic ways to store numbers in most digital computer memories: *fixed point* and *floating point.* (Sometimes these formats are called *integer* and *real*, respectively). The fixed-point format may be used only to represent whole numbers (integers), while the floating-point format may be used to represent either whole numbers or numbers consisting of an integral part and a fractional part.

Most computers are capable of doing calculations in either fixed-point or floating-point formats. The computer user has the responsibility of telling the computer which format or formats it is dealing with. The reason for this is that numbers in either format consist only of zeroes and ones. A computer does not have any way of distinguishing one format from another unless it is told which series of bits represents a floating-point number and which series of bits represents a fixed-point number.

Fixed-Point

The fixed-point format is easier to understand, so let us take that up first. For this discussion assume that the computer locations are 32 bits long (4 bytes). Consider this number:

$$00000000000000000000100010011101_2$$

This is a pure base 2 number and may be converted to decimal in the same way that you converted from binary to decimal in Chapter 2. Let us use base 16 as a stepping stone.

Mark off the number in groups of four digits to obtain the base 16 digits expressing the same number:

$$\frac{0000}{0}\ \frac{0000}{0}\ \frac{0000}{0}\ \frac{0000}{0}\ \frac{0000}{0}\ \frac{1000}{8}\ \frac{1001}{9}\ \frac{1101_2}{D_{16}}$$

Now compute the decimal equivalent of $0000089D_{16}$ by making these calculations:

8 $\times$ 256 (16^2) = 2048

9 $\times$ 16 (16^1) = 144

13(D) $\times$ 1 (16^0) = 13

Sum = 2205_{10} $(0000089D_{16} = 2205_{10})$

Verify this result by either converting the number directly to decimal using powers of 2 or by converting the number to decimal using base 8 as a stepping stone.

Computers must represent negative numbers as well as positive. One way that a computer stores negative numbers is by using the leftmost digit of a binary number to represent whether the number is positive or negative; zero means positive, 1 means negative.

This 32-bit number is understood to be positive:

$$01000001110110111101000001100101_2$$

while this 32-bit number is understood to be negative:

$$11000001110110111101000001100101_2$$

As you can see, the magnitudes of the two numbers are identical but the arithmetic signs are different.

EXERCISE

3-1. Give the 32-bit fixed-point binary equivalents of these base 10 numbers:

(a) 3777_{10}

(b) 8002_{10}

(c) 64_{10}

(d) 123_{10}

(e) 99_{10}

Twos-Complement Arithmetic

Most computers built today employ a more sophisticated and efficient way of representing negative numbers. This method is called the *twos-complement* method.

Here are two numbers having the same magnitude. The first number is positive, the second is negative:

$$000000000000000000001001010010000_2$$

$$1111111111111111111110110101011000_2$$

The decimal value of the first number is 2376, and of the second, -2376.

Exactly how does twos complementing work? For purposes of illustration, we shall use only 6 bits. Keep in mind that what works for 6 bits will work just as well for 32, or more, bits.

Positive numbers are written as pure binary numbers. Thus 001100_2 represents 12_{10} and 010101_2 represents 21_{10}. A negative number is formed by taking its positive form, then changing all the zeroes to ones and all the ones to zeroes, and finally *adding* 1. Let us try 010101_2:

010101_2 the positive number (21_{10})

101010_2 change the zeroes to ones and the ones to zeroes

$\underline{+\ 1_2}$ add 1

101011_2 the negative number (-21_{10})

The positive value 21_{10} is, therefore, 010101_2 and the negative value -21_{10} is 101011_2.

Let us convert 000001 (1_{10}) to its negative form:

000001_2 the positive number (1_{10})

111110_2 change the zeroes to ones and the ones to zeroes

$\underline{+\ 1_2}$ add 1

111111_2 the negative number (-1_{10})

EXERCISE

3-2. Give the 32-bit fixed-point binary equivalents of these base 10 numbers (use twos-complement arithmetic):

(a) -32_{10}

(b) -99_{10}

(c) -123_{10}

(d) -4008_{10}

(e) -3_{10}

6-bit Binary Number	Decimal Equivalent
110001	–15
110010	–14
110011	–13
110100	–12
110101	–11
110110	–10
110111	–9
111000	–8
111001	–7
111010	–6
111011	–5
111100	–4
111101	–3
111110	–2
111111	–1
000000	0
000001	1
000010	2
000011	3
000100	4
000101	5
000110	6
000111	7
001000	8
001001	9
001010	10
001011	11
001100	12
001101	13
001110	14
001111	15

FIG. 3-2

Figure 3-2 shows the entire range of numbers from – 15 through 15 when only 6 bits are used to describe numbers. If these numbers were 32-bit numbers instead of 6-bit ones, there would be 26 additional ones ahead of the negative forms and 26 additional zeroes ahead of the positive forms. The decimal value zero is neither a negative nor positive number. In this method of representing numbers, zero consists of 32 zeroes in a 32-bit computer location.

You can easily determine whether fixed-point numbers are positive or negative. Simply check the leftmost bit. If it is zero, the number is positive; if it is 1, the number is negative.

To determine the magnitude of a fixed-point negative number, use the *identical* procedure for converting a positive number to its negative form. That is, change all the zeroes to ones, all the ones to zeroes, and then add 1. For example, consider this negative number:

$$1111111111111111111011100011001110_2$$

We can easily tell that it is a negative number. (The leftmost bit is 1.) Take the twos complement of the number:

$$000000000000000000100011100011001_2$$

$$+1_2$$

$$\overline{}$$

$$000000000000000000100011100011010_2$$

Now convert the binary number to decimal using base 16 as a stepping stone. The base 16 version of the number is

$$0000239A_{16}$$

Make these calculations:

$$2 \quad \times 4096\,(16^3) = 8192$$

$$3 \quad \times \ \ 256\,(16^3) = \ \ 768$$

$$9 \quad \times \ \ \ 16\,(16^1) = \ \ 144$$

$$10(A) \times \ \ \ \ 1\,(16^0) = \underline{\ \ \ 10}$$

$$\text{Sum} = 9114_{10} \quad (0000239A_{16} = 9114_{10})$$

The decimal value of the binary number is, therefore, -9114_{10}.

There is another way to take the twos complement of a number (positive or negative). Start on the right-hand end of the number and write down whatever zeroes you see there, working toward the left until you see a 1. Copy that 1. Now, from that point moving leftward, change all ones to zeroes and all zeroes to ones. Here is an example:

$$0000101101101011110110000000000000_2$$

Write down the rightmost zeroes:

$$00000000000$$

Now copy the next digit to the left (a 1).

$$100000000000$$

Finally write down the remaining digits of the number but substitute zeroes for ones and ones for zeroes:

$$111101001001010000101000000000000_2$$

The original number, a positive one, has been converted to its negative form. Using the same procedure, we may now recomplement this result and obtain the original number. Try it.

You may use the base 8 or base 16 number systems as stepping stones to the complementing of numbers. Let us try base 8 first. Suppose that the number is

$$0000101101101011110110000000000000_2$$

In octal (base 8), it is expressed as

$$01332754000_8$$

Write down all leftmost zeroes. Then subtract the rightmost nonzero digit from 8. Subtract all remaining digits from 7. These are the steps:

000_8	step 1	copy the rightmost zeroes
4000_8	step 2	subtract the rightmost nonzero digit from 8
76445024000_8	step 3	subtract all other digits from 7

Now write the binary equivalents of the resulting octal digits:

$$111101001001010000101000000000000_2$$

Drop the leftmost 1 to reduce the number of bits to 32 and the task has been accomplished.

Base 16 may also be used. As with base 2 and base 8, leftmost zeroes are copied. The rightmost nonzero digit is subtracted from 16 and all remaining digits are subtracted from 15. The original number:

$$0000101101101011110110000000000000_2$$

Convert to base 16.

$$0B6BD800_{16}$$

00_{16}	step 1	copy the rightmost zeroes
800_{16}	step 2	subtract the rightmost nonzero digit from 16
$F4942800_{16}$	step 3	subtract all other digits from 15

Convert to binary:

$$11110100100101000010100000000000_2$$

Keep in mind that $15 - 0 = 15$ (F), $15 - B$ $(11) = 4$, and $15 - D$ $(13) = 2$.

So that we may study the various facets of fixed-point numbers using twos complementing for negative values, let us examine a complete system involving 5 bits. This discussion is academic, of course, since no real-life computers (except possibly toys) use 5-bit memory locations. But the principles that are valid for 5-bit numbers are also valid for 32-bit systems.

Figure 3-3 shows all possible 5-bit fixed-point numbers. As you can see, the largest positive number is 01111 ($2^4 - 1$) and the smallest negative number is 10000 (-2^4). In magnitude, the smallest negative number is larger by 1 than the largest positive number.

Where a 32-bit number is concerned, the largest positive number consists of a zero followed by 31 ones. This value is $2^{31} - 1$. The smallest negative number consists of a 1 followed by 31 zeroes. This value is -2^{31}. Again, in magnitude, the smallest negative number is larger by 1 than the largest positive number.

Now refer to the 5-bit spectrum. How does a positive value change if the leftmost digit is erased and zero is added to the right.

5-bit Number	Decimal Equivalent
10000	−16
10001	−15
10010	−14
10011	−13
10100	−12
10101	−11
10110	−10
10111	−9
11000	−8
11001	−7
11010	−6
11011	−5
11100	−4
11101	−3
11110	−2
11111	−1
00000	0
00001	1
00010	2
00011	3
00100	4
00101	5
00110	6
00111	7
01000	8
01001	9
01010	10
01011	11
01100	12
01101	13
01110	14
01111	15

FIG. 3-3

Examples:

$$00001_2 \text{ changes to } 00010_2$$
$$00010_2 \text{ changes to } 00100_2$$
$$00011_2 \text{ changes to } 00110_2$$

The value doubles. In a 32-bit number, the same principle may be applied. That is, the value

$$00000000000010000000011000110000_2$$

may be doubled by dropping the leftmost digit and adding zero at the right:

$$0000000000010000000011000110000_2$$

The value may be multiplied by 4 by adding two zeroes, by 8 by adding three zeroes, by 16 by adding four zeroes, etc.

The fact that values may be multiplied by some power of 2 is a useful one; computer programmers employ the principle whenever it is expedient to do so.

Sometimes a number permits no further doubling. Consider this number taken from the 5-bit spectrum:

$$01010_2$$

This value cannot be doubled because to do so would exceed the range of the positive side of the number range. As you know, the maximum number in the 5-bit range is 01111_2. Doubling 01010_2 would cause it to appear as 10100_2, which is not the double of 01010_2 but a *negative* number. Which?

In real-life computers, any attempt by a user to exceed the maximum range of a number, either positive or negative, will cause error conditions to arise. Sometimes these error conditions cause error messages to be printed, sometimes not.

What happens to a *negative* number if the leftmost 1 is dropped and a zero is added to the right? The result is a doubling of the number in the negative direction. That is, 11111_2 (-1) becomes 11110_2 (-2) and 11110_2 (-2) becomes 11100_2 (-4), etc. As with positive numbers, the process of doubling a negative number cannot be continued beyond the range of the number system. (In the positive direction, the range is exceeded when leftmost zeroes run out, and in the negative direction, the range is exceeded when leftmost ones run out.)

What is the result when positive and negative values are added. Let us use the 5-bit spectrum to add 2 to -4.

$$00010_2$$
$$+ \ 11100_2$$
$$\overline{11110_2}$$

The answer is -2, as it should be. Numbers must not exceed their maximum ranges, of course. In a 5-bit system we may not add

$$10000_2 \quad (-16)$$
$$+\,11110_2 \quad (-2)$$

because, in magnitude, the result will exceed the maximum permissible value in the negative direction. Nor may we add

$$01000_2 \quad (8)$$
$$+\,01010_2 \quad (10)$$

because of range problems.

Subtraction in twos-complement arithmetic is easy. Suppose that you need to subtract 3 from 11. Change the problem from a subtraction problem to an addition problem by complementing the subtrahend (3) and *adding* it to the minuend (11). Example:

$$
\left.
\begin{array}{r}
01011_2 \\
\text{Subtract } \underline{00011_2}
\end{array}
\right\}
\text{The original problem}
$$

$$
\left.
\begin{array}{r}
\text{Change to } 01011_2 \\
\text{Add } \underline{11101_2} \\
101000_2
\end{array}
\right\}
\begin{array}{l}
\text{changed to a different but} \\
\text{equivalent form}
\end{array}
$$

Drop the leftmost bit since a 5-bit system cannot hold 6 bits. The answer is 01000_2, which is, of course, 8. Let us try another problem: subtract -5 from -2:

$$
\left.
\begin{array}{r}
11110_2 \\
\text{Subtract } \underline{11011_2}
\end{array}
\right\}
\text{the original problem}
$$

$$
\left.
\begin{array}{r}
\text{Change to } 11110_2 \\
\text{Add } \underline{00101_2} \\
100011_2
\end{array}
\right\}
\begin{array}{l}
\text{changed to a different but} \\
\text{equivalent form}
\end{array}
$$

Dropping the leftmost bit, we find that the answer is 00011_2. This is the correct answer.

EXERCISE

3-3. (a) Add these values:

$$00001011_2$$
$$00010011_2$$
_____

(b) Add these values:

$$00010000_2$$
$$00111101_2$$
_____

(c) Subtract the bottom number from the top one:

$$00011011_2$$
$$00001101_2$$
_____

(d) Subtract the bottom number from the top one (give the answer in 8 bits):

$$00011100_2$$
$$11111001_2 \quad \text{(a negative value)}$$
_____

(e) Subtract the bottom number from the top one (give the answer in 8 bits):

$$11100011_2 \quad \text{(a negative value)}$$
$$11110001_2 \quad \text{(a negative value)}$$
_____

(f) Subtract the bottom number from the top one (give the answer in 8 bits):

$$11111100_2 \quad \text{(a negative value)}$$
$$00001101_2$$
_____

Now let us consider multiplication. For this discussion we shall expand the length of binary numbers to 10 bits. The range of numbers that may be expressed in a 10-bit spectrum is from -2^9 (-512) through $2^9 - 1$ (511). (The smallest number is 1000000000_2 and the largest number is 0111111111_2.)

In the first example, we shall multiply 0000010010_2 by 0000010110_2. Since the leftmost digit in each number is zero, both numbers are positive and the result will be positive. We can work out the solution like this:

$$0000010010_2$$
$$0000010110_2$$

$$0000000000$$
$$0000010010$$
$$0000010010$$
$$0000000000$$
$$0000010010$$
$$0000000000$$
$$0000000000$$
$$0000000000$$
$$0000000000$$
$$0000000000$$

$$0000000000110001100_2$$

To check the answer, we convert 0000010010_2 and 0000010110_2 to decimal, multiply them, and check whether the product is consistent with 0000000000110001100_2. The original values are 18_{10} and 22_{10}, respectively, and the product is 396_{10}. The result checks.

Observe that the answer must not exceed the maximum range allowed by a 10-bit number. Since 396_{10} is less than 511_{10}, this restraint is observed.

The computer does not work out a binary multiplication problem the same way we did it above. It uses a process that involves shifting and adding. We have already discussed shifting, but let us review it here. A number is shifted left by dropping the leftmost bit and adding zero at the right. Thus

$$0000010001_2 \quad (17_{10})$$

when shifted left once gives

$$0000100010_2 \quad (34_{10})$$

If the number is shifted left twice more, its new representation becomes

$$0010001000_2 \quad (136_{10})$$

You will recall that shifting a number left one place doubles it; shifting it left two places multiplies it by 4; three places multiplies it by 8; etc.

Now let us examine how a computer multiplies 0000010010_2 by 0000010110_2. We shall call the first number the multiplicand and the second number the multiplier. We need a register to receive the product. Let us call that register the *product register* and initialize it with ten binary zeroes:

$$0000000000_2$$

Now scan the multiplier from left to right until a nonzero bit is found. (If no nonzero bit is found, no further action is required. The answer to the problem is 0000000000_2.) The sixth bit (counting from the left) in the multiplier is a 1. *Having found* a nonzero bit in the multiplier, we add the *multiplicand* to the *product register*. The product register now looks like this:

$$0000010010_2$$

This is the rule that applies: whenever a zero is found in the multiplier, the product register is shifted left one place. Whenever a 1 is found in the multiplier, the product register is shifted left one place *and the multiplicand is added to the product register.*

Continue to scan the multiplier from left to right. The seventh bit (counting from the left) in the multiplier is zero; therefore the product register is shifted left one place. It becomes

$$0000100100_2$$

The eighth bit (still counting from the left) in the multiplier is 1; therefore the product register is shifted left one place *and* the multiplicand is added to the register. These two actions cause this:

$$0001001000_2$$
$$+\ 0000010010_2$$
$$\overline{0001011010_2}$$

Note that the multiplicand does not change.

The ninth bit in the multiplier is also a 1; therefore the product register is shifted left one place:

$$0010110100_2$$

and the multiplicand is added:

$$0010110100_2$$
$$+\ 0000010010_2$$
$$\overline{0011000110_2}$$

Finally, the tenth bit, being a zero, causes a final left shift of the product register:

$$0110001100_2$$

This is the product of $0000010010_2 \times 0000010110_2$.

Now, to check our work, let us reverse the roles of the multiplicand and multiplier. The number 0000010110_2 will be the multiplicand, and the number 0000010010_2 will be the multiplier. Here are the original and subsequent states of the product register as bits of the multiplier are checked.

0000000000_2	original state
0000010110_2	sixth bit of multiplier is 1
0000101100_2	seventh bit of multiplier is zero
0001011000_2	eighth bit of multiplier is zero
0010110000_2	ninth bit of multiplier is 1
$+\ 0000010110_2$	
0011000110_2	
0110001100_2	tenth bit of multiplier is zero

The result is, of course, the same as it was before.

When negative numbers are involved, the computer must adjust the sign of the product. If a negative number is to be multiplied by a positive number, the negative number is changed to its positive form, the product is obtained, and the result is changed to a negative number; if two negative numbers are to be multiplied, both numbers are changed to their positive forms and the product is obtained and retained as a positive number.

Fixed-point division may easily be performed by a computer in much the same way that it does multiplications. We shall not go into a detailed explanation of the process here but will give some examples so that if you wish to do some independent work, you may.

Problem: Divide 0001101000_2 (104_{10}) by 0000001000_2 (8_{10}). The first value is the *dividend* and the second value is the *divisor*. The dividend is placed in the working register. The result (quotient) builds up in a register called the *quotient register*.

	Working Register	Quotient Register
Dividend	0001101000_2	0000000000_2
Subtract divisor (shifted to match dividend)	0001000000_2	0000000001_2
New dividend	0000101000_2	
Subtract divisor (shifted one more place)	0000100000_2	0000000011_2
New dividend	0000001000_2	
Subtract divisor (shifted two more places	0000001000_2	0000001101_2
New Dividend	0000000000_2	

The last value in the quotient register gives the quotient 0000001101_2 with no remainder.

Problem: Divide 0110110001_2 (433_{10}) by 0000010101_2 (21_{10}).

Dividend	0110110001_2	0000000000_2
Subtract Divisor (shifted to match dividend)	0101010000_2	0000000001_2
New dividend	0001100001_2	
Subtract divisor (shifted two more places)	0001010100_2	0000000101_2
New dividend	0000001101_2	
Divisor shifted two more places	0000010101_2	0000010100_2

The quotient is 10100_2 (20_{10}) with 1101_2 (13_{10}) remainder. (Note that if 0000010101_2 is shifted any more, binary digits at the right of the number begin to be dropped. Division is terminated with the quotient register giving a portion of the answer and the remainder is obtained from the working register.)

Problem: Divide 0010101110_2 (174_{10}) by 0000000110_2 (6_{10}).

	Working Register	Quotient Register
Dividend	0010101110_2	
Subtract divisior (shifted to match dividend)	0001100000_2	0000000001_2
New dividend	0001001110_2	
Subtract divisor (shifted one more place)	0000110000_2	0000000011_2
New dividend	0000011110_2	
Subtract divisior (shifted one more place)	0000011000_2	0000000111_2
New dividend	0000000110_2	
Subtract divisor (shifted two more places)	0000000110_2	0000011101_2
New dividend	0000000000_2	

The quotient is 11101_2 (29_{10}) with no remainder.

Earlier in this chapter, we said that where a computer is concerned there are two ways that it stores numbers in its memory: in fixed-point (integer) and floating-point (real). We have completed a discussion of fixed-point arithmetic and will turn now to floating-point arithmetic.

Fixed-point formats enable us to express integers (whole numbers) only. We would like the ability to deal with numbers containing integer parts and fractional parts. Many calculations involve numbers such as 3.25, .0094, and −3.8, for instance. Floating-point formats give us the ability to use such numbers.

Floating Point

A floating-point number consists of two parts, the *mantissa* and the *characteristic*. To refresh our memory as to how mantissas and characteristics work, let us look at some examples in base 10. The value 3126.34 is equivalent to

$$.312634 \times 10^4$$

$$\text{or} \quad 3.12634 \times 10^3$$

$$\text{or} \quad 31.2634 \times 10^2$$

$$\text{or} \quad 312634. \times 10^{-2} \quad \text{etc.}$$

There are many ways to express 3126.34 when you express it as a fraction times some power of 10. We are used to dealing in numbers expressed in base 10, so much of what we have said so far may seem obvious. But now let us consider a computer that deals in base 2. The value 4.5 could be expressed as $.5625 \times 2^3$, $.28125 \times 2^4$, 1.1250×2^2, etc. Keep in mind that the last sentence was written using base 10 numbers but that the memory of a computer uses only zeroes and ones (base 2) to express these values.

In floating point, a number is expressed as a fraction multiplied by some power of 2. The fraction is the mantissa and the power of 2 is the characteristic. There are several floating-point systems in use; we shall discuss two. The first we shall call the *full-range system*; the second, the *excess-64 system*. Most systems you encounter will use some form of these two basic systems.

Full-Range Floating-Point System

Consider a computer that contains several thousand 36-bit memory locations. Any location may contain a floating-point number. If it does, the number has two parts, the mantissa and the characteristic. Figure 3-4 shows such a memory location.

For positive numbers the mantissa is always a fraction having a value of at least .5 but not reaching 1. Thus the value of a mantissa may be .625, .5, .8999, etc., but it may never be as large as 1. For negative numbers, the mantissa must be equal to or algebraically less than -.5. The smallest possible mantissa may be exactly -1.

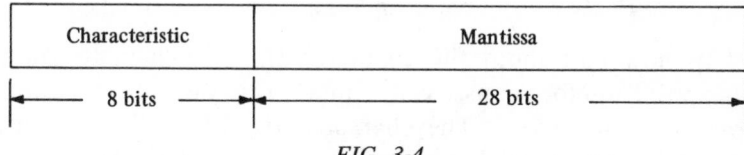

FIG. 3-4

Figure 3-5 illustrates a computer location containing a mantissa expressing the value .625. The leftmost bit of mantissa tells whether the number is negative or positive (1 means negative and zero means positive). A decimal point is assumed to exist between the sign indicator and the second bit of the mantissa. The bits that follow the assumed decimal point have positional values of .5 (2^{-1}), .25 (2^{-2}), .125 (2^{-3}), etc. In the example, the value of the mantissa is .625 because the sign indicator tells that the number is positive and because there is 1 bit in the positions representing .5 and .125. These values are summed to obtain the value of the mantissa.

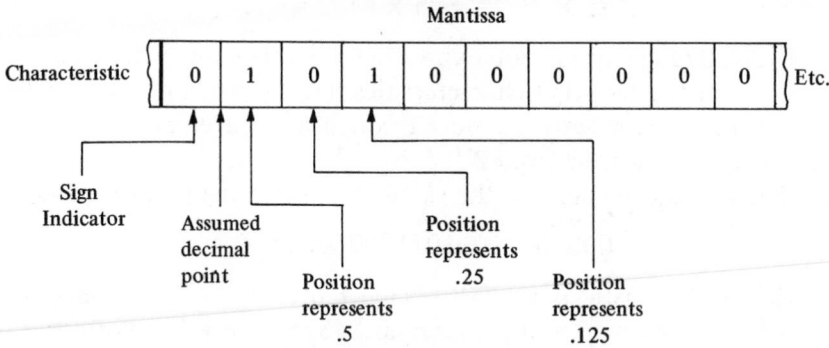

FIG. 3-5

Mantissas are expressed in twos-complement arithmetic. Therefore the mantissa -6.25 would be expressed as shown in Fig. 3-6. Whenever faced with a negative mantissa to evaluate, convert the mantissa to its positive form and then add a minus sign. From the leftmost bit, you can easily tell that the above mantissa is negative.

Mantissa

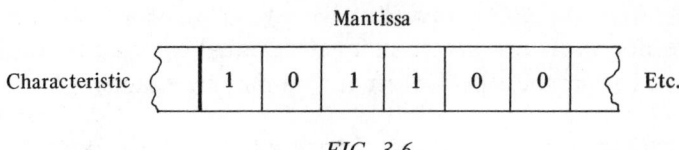

FIG. 3-6

Let us now talk about the characteristic. The characteristic in a floating-point number consists of 8 bits. (This varies from computer to computer, of course.) The characteristic (Fig. 3-7) represents a whole-number power of 2. *Evaluate the value of the characteristic as an integer.* The value of the characteristic is 13. Therefore it represents 2^{13}. The 2 is *understood*.

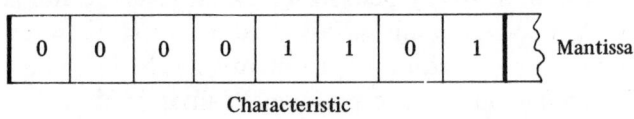

FIG. 3-7

You can see that if a characteristic is 13 and a mantissa is .625, the number represented is

$$.625 \times 2^{13} = .625 \times 8192 = 5120_{10}$$

The full-range system of expressing floating-point values gets its name from the fact that characteristics run through a range of values from some given negative number through some given positive one. A typical range might be from 2^{-128} to 2^{127}.

See if you can compute the value of this floating-point number:

$$00000011011011000000 \ldots 0_2$$

The characteristic is 00000011 and represents 3. The mantissa is 011011000000 . . . 0 and represents .84375. The value expressed is therefore

$$.84375 \times 2^3 = .84375 \times 8 = 6.75_{10}$$

Characteristics may be negative. Consider this next one:

$$11110111_2$$

The leftmost bit, being 1, indicates that the value of the characteristic is negative. By converting to the positive form, we find that the characteristic represents -9. Now if the mantissa is .625, the value being represented is

$$.625 \times 2^{-9} = .625 \times .001953125 = .001220703125_{10}$$

The number is a positive number despite the fact that the characteristic is negative. Negative characteristics simply represent values whose magnitudes lie between zero and 1. Those values may, of course, be either positive or negative.

When 8 bits are employed in a characteristic and twos-complement arithmetic is being used, the largest positive characteristic is $2^7 - 1$ (127) and the largest negative characteristic is 2^7 (128). When 28 bits are used to express the mantissa the maximum number of significant digits known with certainty about any value is 8. Therefore it is possible for a person to receive an answer that indicates that a very large or a very small number is involved ($\pm 1 \times 10^{\pm38}$, for example), yet that number may be expressed so that only the first eight digits are known with certainty.

Let us convert some decimal values to floating point. The first is 92.4_{10}. To determine what the exponent must be, we look for the smallest power of 2 that gives a value *larger* than 92.4. That power is 7. (Note that $2^6 = 64$; $2^7 = 128$; $2^8 = 256$.) We have found the power of 2 to use in the characteristic. The characteristic is therefore

$$00000111_2$$

Now, to obtain the mantissa, divide 92.4 by 128 ($2^7 = 128$). The fraction that results is .721875. Observe that this fraction is equal to or greater than .5 but is not as large as 1. We obtained this fraction and no other because we divided by a number that guarantees a fraction within the prescribed limits. If we had divided by 64 (2^6), the result would have been larger than 1 (too large), and if we had divided by 256 (2^8), the fraction would have been smaller than .5 (too small).

Having obtained the fraction required for the mantissa, we must now convert it to binary. You already know how to convert decimal fractions to binary, so the work in Fig. 3-8 should be clear. The fraction required for the mantissa is $.5614631463_8$. Ten octal digits represent 30 bits but we can use only 28. Also, we need a zero bit ahead of the fraction to indicate that the number is positive. The 28 bits actually used in the mantissa therefore are

Spillovers
give answer ⟶

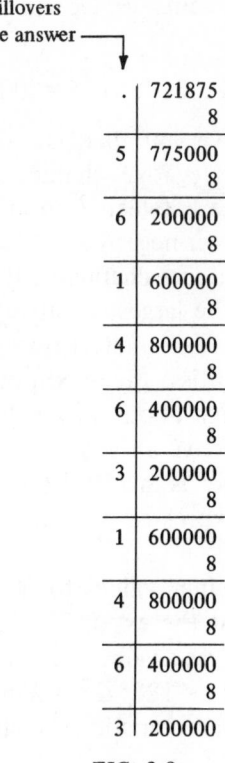

.	721875
	8
5	775000
	8
6	200000
	8
1	600000
	8
4	800000
	8
6	400000
	8
3	200000
	8
1	600000
	8
4	800000
	8
6	400000
	8
3	200000

FIG. 3-8

$$\underset{.5 \quad 6 \quad 1 \quad 4 \quad 6 \quad 3 \quad 1 \quad 4 \quad 6_8}{0101 \ 110 \ 001 \ 100 \ 110 \ 011 \ 001 \ 100 \ 110_2}$$

As you can see, this binary fraction does not exactly equal .721875, but it is as close as we can get. Putting characteristic and mantissa together, we find that the floating-point representation of 92.4_{10} is

$$000001110101110001100110011001100110_2$$

Expressed octally, this number is

$$016561463146_8$$

Now let us work out a problem. Suppose that we are given this number to convert from floating point to decimal:

$$777400000000_8$$

The first step is to expand the number and isolate its characteristic and mantissa parts:

<u>1 1 1 1 1 1 1 1</u> <u>1 1 0000000000000000000000000000</u>

Characteristic   Mantissa

Both characteristic and mantissa values are negative. By converting both numbers to their positive forms, we find that the number shown is

$$-.5 \times 2^{-1} \quad \text{or} \quad -.5 \times .5 = -.25$$

In the full-range system, the value zero has a special representation. It is 40000000000_8. In binary, this is a 1 bit followed by 35 zeroes.

When you instruct a computer to solve problems, you must tell it whether the numbers you are using are fixed-point or floating-point. If you command the computer to work with floating-point values, it will add, subtract, multiply, and divide them and then *adjust the results* to agree with floating-point formats.

Suppose, for example, that you command the computer to multiply 4 × 16. In memory, the values are stored as

$$.5 \times 2^3 \quad \text{and} \quad .5 \times 2^5$$

The computer will multiply .5 by .5, giving .25, and add the exponents, giving 2^8. Now the result is $.25 \times 2^8$. The answer is correct, but it is not in normal form, that is, where the mantissa is equal to or greater than .5. The computer *normalizes* the product by doubling .25 (giving .5) and reducing the power of 2 by 1 (giving 2^7). The answer has been *normalized* and is stored as $.5 \times 2^7$.

We said that there are two basic ways to form floating-point numbers: the full-range system and the excess-64 system. We have just described the full-range system, which has an 8-bit characteristic and 28-bit mantissa. The principles would be the same if some system you encounter has a 10-, 12-, 16-, or 20-bit characteristic, etc., or if it has a 32-, 36-, 48-, or 64-bit mantissa, etc.

Excess-64 Floating-Point System

The excess-64 system is a system in which the characteristic varies from zero through 127. If a number has a characteristic that lies between zero and 63, inclusive, the characteristic is adjusted so that it becomes a negative characteristic between -64 and -1; if the characteristic lies between 64 and 127, inclusive, it is adjusted so that it lies between zero and 63.

The adjustment is made by subtracting 64 from the characteristic.

Characteristics are expressed in terms of base 16. If the integer value of the characteristic is 67, for example, 64 is subtracted. The remainder (3) is applied as the power of 16. The effective characteristic is therefore 16^3.

Here is a table showing nominal to effective exponent conversions:

Nominal Power of 16	Effective Power of 16
0	-64
1	-63
2	-62
.	.
.	.
.	.
62	-2
63	-1
64	0
65	1
66	2
.	.
.	.
.	.
126	62
127	63

Given a nominal characteristic, you may compute its effective value by subtracting 64. Thus, if the nominal characteristic is 70, the effective characteristic is $70 - 64 = 6$; if the nominal characteristic is

38, the effective characteristic is $38 - 64 = -26$. In the former example, the value of the characteristic is 16^6, and in the latter, 16^{-26}

The system we describe here is one used on IBM-360 and IBM-370 computers and is based upon excess-64, but the principles are the same whether the system you encounter is excess-50, -100, -128, etc. In this system, the size of the mantissa is 24 bits, but the principles are the same if the mantissa in the system you encounter has 30, 36, 48, 64 bits, etc.

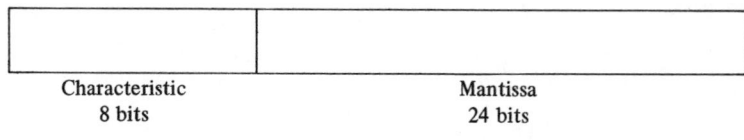

| Characteristic
8 bits | Mantissa
24 bits |

FIG. 3-9

Consider the 32-bit floating-point number in Fig. 3-9. Let us consider the characteristic first. The normal way to express an IBM-360/370 floating-point characteristic is in base 16 (hexadecimal) digits. Consider, for example,

$$2C_{16}$$

The bits represented are 00101100_2. The value of the binary number, expressed as an integer, is 44_{10}, and the characteristic being represented is $44 - 64 = -20$. Therefore, to obtain the value of the floating-point number, the mantissa has to be multiplied by 16^{-20}.

The leftmost bit of the characteristic tells whether the floating-point number itself is positive or negative: zero means positive, 1 means negative. Therefore the value of a floating-point number that has the characteristic $2C_{16}$ is positive (because the leftmost bit in 00101100_2 is zero).

It is simple to determine whether a floating-point number is positive or negative. If the leftmost hexadecimal digit lies between zero and 7, inclusive (0, 1, 2, 3, 4, 5, 6, 7), the number is positive; if it lies between 8 and F, inclusive (8, 9, A, B, C, D, E, F), the number is negative.

The excess-64 system *does not* use twos-complement arithmetic. To show that the floating-point value itself is negative, the leftmost bit of the characteristic is 1; to show that the characteristic is negative, the value of the characteristic is less than 64.

Now let us look at mantissas. Consider this mantissa:

$$3B0000_{16}$$

In binary this value is

$$0011101100000000000000000_2$$

The positional values of the bits are 2^{-1} (.5) for the leftmost bit, 2^{-2} (.25) for the next bit to the right, etc. The value of the mantissa shown here is therefore

$$
\begin{aligned}
0 \times .5   &= 0 \\
0 \times .25   &= 0 \\
1 \times .125   &= .125 \\
1 \times .0625   &= .0625 \\
1 \times .03125   &= .03125 \\
0 \times .015625   &= 0 \\
1 \times .0078125 &= .0078125 \\
1 \times .00390625 &= \underline{.00390625} \\
\text{Sum}   &= .23046875_{10}
\end{aligned}
$$

The value of the mantissa $3B0000_{16}$ is, therefore, $.23046875_{10}$. This value, when multiplied by the characteristic of the floating-point number, gives its value.

In the excess-16 system, floating-point zero is represented by a string of zeroes; in a 32-bit number, by 32 zeroes.

Now let us evaluate some floating-point numbers:

$$41A00000_{16}$$

This number consists of the characteristic 41_{16} and the mantissa $A00000_{16}$. The number is positive because the leftmost digit of the characteristic is 4_{16} (0100_2).

The nominal value of the characteristic is computed as follows:

$$4 \times 16 \, (16^1) = 64$$
$$1 \times 1 \, (16^0) = \underline{1}$$
$$\text{Sum} = 65_{10}$$

Now, $65_{10} - 64_{10} = 1_{10}$; therefore the effective characteristic is $16^1 = 16_{10}$.

The mantissa is computed using the positional values shown below:

Mantissa	A	0	0	0	0	0_{16}
Positional values	16^{-1}	16^{-2}	16^{-3}	16^{-4}	16^{-5}	16^{-6}

$$10(A) \times 16^{-1} = .625$$
$$0 \quad \times 16^{-2} = 0$$
$$0 \quad \times 16^{-3} = 0$$
$$0 \quad \times 16^{-4} = 0$$
$$0 \quad \times 16^{-5} = 0$$
$$0 \quad \times 16^{-6} = \underline{0}$$
$$\text{Sum} = .625_{10}$$

The decimal value of 16^{-1} is, of course, $\frac{1}{16}$ or .0625. Having computed the value of the characteristic (16_{10}) and the value of the mantissa ($.625_{10}$), we can now compute the decimal equivalent of the floating-point value $41A00000_{16}$. It is $16_{10} \times .625_{10}$ or 10_{10}.

Let us try a more complicated value:

$$C33CE000_{16}$$

The number being represented is negative. We can determine this from the leftmost of the two characteristic characters, $C3_{16}$. (As you know, C_{16} corresponds to 1100_2.)

Obtain the decimal equivalent of $C3_{16}$ this way:

$$4 \times 16 \, (16^1) = 64$$
$$3 \times 1 \, (16^0) = \underline{3}$$
$$\text{Sum} = 67_{10}$$

Subtract 64. The result is 3 and the effective characteristic is 16^3 or 4096_{10}. In the above calculation, 4_{16} was used as the multiplier for 16^1, not C_{16}, because the leftmost bit of C_{16} is not truly a part of the characteristic. It simply tells that the floating-point value is negative. When computing a characteristic, that bit must be ignored or treated the same as if it were zero.

Now we may evaluate the mantissa $3CE000_{16}$.

3     $\times$ .0625 $(16^{-1}) = .1875$

12(C) $\times$ .00390625 $(16^{-2}) = .046875$

14(E) $\times$ .000244140625 $(16^{-3}) = .00341796875$

$$\text{Sum} = .23779296875_{10}$$

This mantissa, .23779296875, multiplied by the value of the characteristic, 4096_{10}, gives the value of the floating-point number, exactly 974_{10}. But recall that the value is negative. Therefore $C33CE000_{16} = -974_{10}$.

When you need to convert a decimal number to excess-64 floating point, you use a method that is similar to the method used in converting a decimal number to full-range floating point.

Suppose that you need to convert 262.144_{10} to floating point, for example. Select a power of 16 that gives a value *larger* than 262.144_{10}. This power is 3, because $16^2 = 256_{10}$ (too small) and $16^4 = 65536_{10}$ (too large). Now we add 64_{10} to the power, giving 67_{10}. Converting 67_{10} to hexadecimal, we get 43_{16}. This is the characteristic. To form the mantissa, we divide 16^3 (4096_{10}) into; 262.144_{10}. This gives the fraction $.064_{10}$. Now we must convert the fraction to base 16 (Fig. 3-10). The mantissa's value is $10624D_{16}$. (Note that the hexadecimal values are shown in boldface type.)

We now join the characteristic to the mantissa and get $4310624D_{16}$. This is the floating-point representation of 262.144_{10}. If the value to be converted had been -262.144_{10}, the only change needed in the floating-point number would have been in the leftmost digit: the value 4_{16} would have been C_{16}. The value -262144_{10} would therefore be written in floating-point as $C310624D_{16}$.

Here is another example: convert $.01_{10}$ to floating-point. First

```
Spillovers give  ───┐
answer              ↓
                  .  | 064
                       16
                    ─────────
                 0 | 384
                     64
                    ─────────
                 1 | 024
                     16
                    ─────────
                 0 | 144
                     24
                    ─────────
                 0 | 384
                     16
                    ─────────
                 2 | 304
                 3 | 84
                    ─────────
                 6 | 144
                     16
                    ─────────
                 1 | 864
                     44
                    ─────────
                 2 | 304
                     16
                    ─────────
                 1 | 824
                 3 | 04
                    ─────────
                 4 | 864
                     16
                    ─────────
                 5 | 184
                 8 | 64
                    ─────────
                13 | 824
```

FIG. 3-10

we find that the power to use in the characteristic must be -1, because $16^0 = 1$ (too large) and $16^{-2} = .00390625$ (too small). Add 64_{10} to -1_{10}. This gives 63_{10}, the value of the characteristic. This number in binary looks like this:

$$00111111_2$$

which is $3F_{16}$ in hexadecimal. Now divide $16^{-1}(.0625_{10})$ into $.01_{10}$. This gives the fraction $.16_{10}$. Next convert the fraction to hexadecimal (Fig. 3-11). The hexadecimal version of $.01_{10}$'s mantissa is $28F5C2_{16}$. Putting characteristic and mantissa together, we find that the floating-point version of $.01_{10}$ is $3F28F5C2_{16}$.

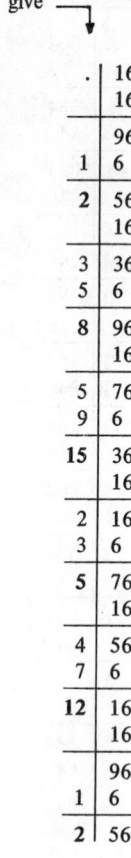

FIG. 3-11

This method of converting a decimal number to floating-point will always give a *normalized* number. That is, the mantissa will always be between .0625 and 1.

If the task of converting from a decimal value to floating-point has seemed tedious, do not despair; the computer makes these conversions when you command it to do so. Nevertheless, you should know what the form is so that you will be able to convert a number by hand in certain instances when you need to debug computer programs.

EXERCISES

3-4. How much is a K of memory? (Assume that the computer is byte-oriented and identifies each byte with an address.)

3-5. How much is a K of memory if the computer is "word"-oriented?

3-6. Define *bit*.

3-7. Define *byte*.

3-8. What are the two basic formats by which numbers may be stored in the memory of a computer?

3-9. What is the value of the following fixed-point 32-bit number:

$$00000000000000010000010000110001_2$$

3-10. Express the following 32-bit number as a base 16 number:

$$00101110110100000101111000000110_2$$

3-11. Express the binary number found in Exercise 3-10 as a base 8 number.

3-12. Expressed as a binary number, what is the negative version of the binary number found in Exercise 3-10?

3-13. What is the positive version of the following 32-bit number:

$$11111111110011111100101011000011_2$$

3-14. Expressed as a decimal value, what is the value of the following binary number:

$$11111111111111111111111111111111_2$$

3-15. Assuming that a computer you are working with uses only 4 bits to express numbers, construct a chart showing all possible 4-bit values, both negative and positive.

3-16. In the chart you made in response to Exercise 3-15, what is the largest positive value that the system can express?

3-17. In the chart you made in response to Exercise 3-15, what is the largest negative value that the system can express?

3-18. Multiply 0000010110 by 0000001101 the way a computer would do it. Show the product register as it develops the product.

3-19. Name the two parts that are used to form a floating-point number.

3-20. Name the two floating-point formats discussed in the text.

3-21. What is the binary value of the characteristic in the full-range floating-point system when the value 28.5 is being expressed? (Assume that the characteristic has an 8-bit form.)

3-22. What is the binary value of the characteristic in the full-range floating-point system when the value .125 is being expressed? (Assume that the characteristic has an 8-bit form.)

3-23. Study the following characteristics. Tell which ones indicate that the entire value being expressed is negative: (Excess-64 system)

$$41_{16}$$

$$C3_{16}$$

$$A4_{16}$$

$$FA_{16}$$

$$2F_{16}$$

3-24. Using the excess-64 floating-point system determine the decimal equivalent of

$$B33C0000_{16}$$

3-25. Using the excess-64 floating-point system, determine what the mantissa must be to describe the value 64. Express the mantissa as six base 16 characters.

3-26. What is the excess-16 floating-point format of the value 3.6? Use eight base 16 characters to describe the result: two characters for the characteristic and six characters for the mantissa.

Chapter 4

ELEMENTARY COMPUTER PROGRAMMING

A computer has been compared to an unimaginative clerk, to a giant desk calculator, and to an electronic brain. A computer is none of these things; it is simply a machine that processes information. The computer's strength is not so much that it can do wonderful things—rather, it is the fact that it does things exceedingly fast and with great accuracy.

One may think of a computer system as shown in Fig. 4-1, which shows that information, whether it be in the form of numbers, such as 3.91, −8.2, and 15.4, or words such as TEMPERATURE, JONES, and ALASKA, is manipulated, moved about, changed, and then printed out in easily understandable form.

A user may wish to give a computer the numbers 8.78 and 2.74 and instruct it to add them, then add a dollar sign, and print the result, $11.52, on a check. Or, he may wish to give the computer the name A W JONES and instruct it to add some punctuation, rearrange the various parts, and print it in the form JONES, A.W. in a company phone book. Or a user may give a computer some information about the current demands on a power-generating plant and instruct it to decide whether or not additional power must be generated *right now* to satisfy customer requirements.

Input to a computer system may come from a variety of sources. Output depends on what is required. Figure 4-2 shows several ways that information may enter a system and what output may be generated.

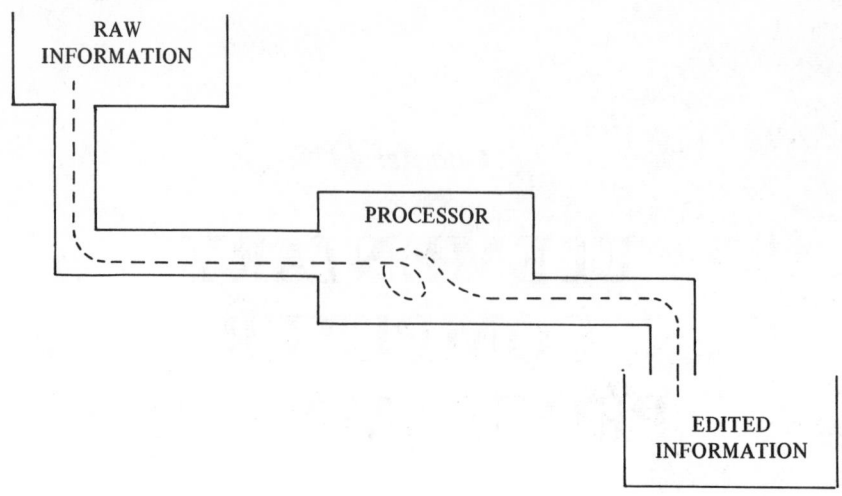

FIG. 4-1

Computer systems may be called upon to work upon hundreds, if not thousands, of different tasks in a single day. Such systems are called *general-purpose* systems because they are not restricted in the kinds of jobs that may be assigned to them. Opposed to the general-purpose systems are the *dedicated systems,*which have only one or two reasons for existing: to control a steel mill, for example, or to guide a spaceship to the moon.

Figure 4-2 shows that input information may come from punched cards, from magnetic tapes, or directly from an operating piece of machinery. Input may also come from devices not shown: paper tape, disk, optical character readers, and others.

In this text, it is not our purpose to go deeply into a technical description of the construction or operation of computers. But we do want to describe an almost universal method of getting information into the memory of a computer—through the use of punched cards.

Before we look at a punched card, let us consider what a computer's memory is like. A computer's memory may be thought of as a long string of pigeonholes, each hole capable of holding a digit, a letter of the alphabet, or a special sign (Fig. 4-3). There may be many thousands of these pigeonholes in the string. The computer has the ability to store in a pigeonhole any character required by the

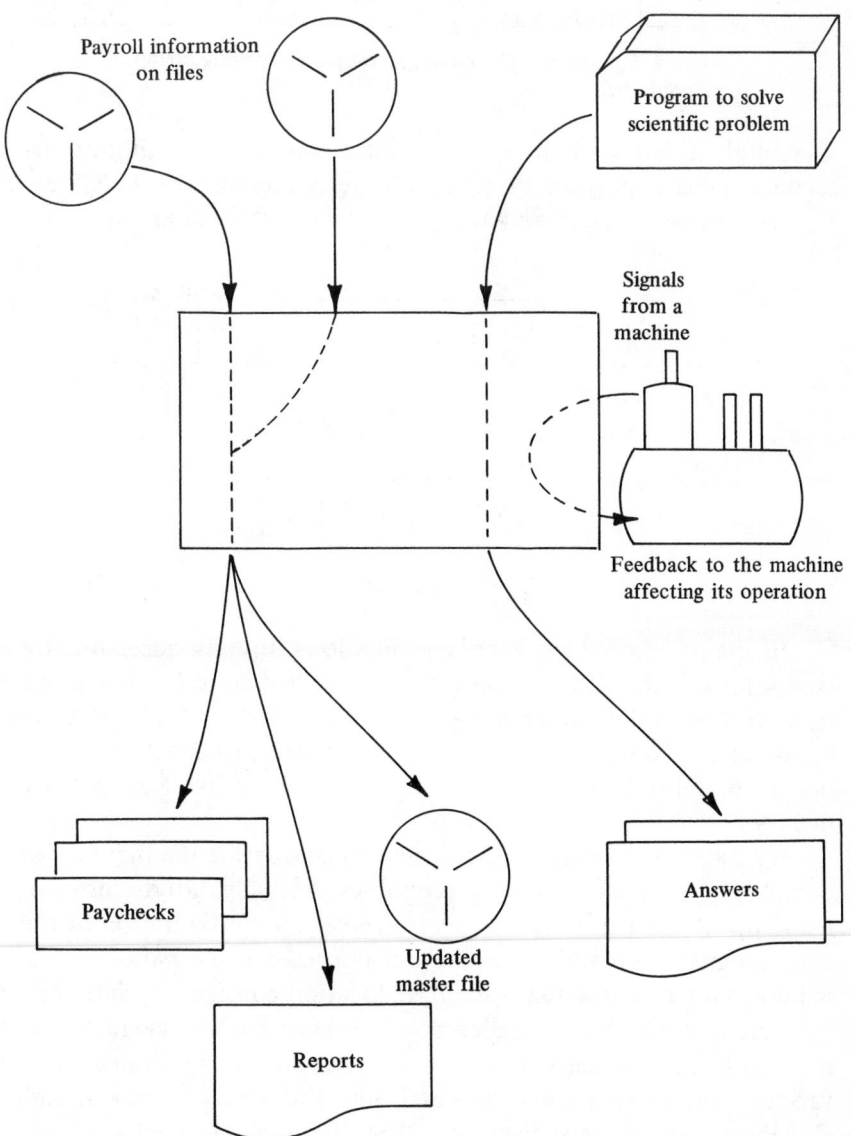

Payroll information on files

Program to solve scientific problem

Signals from a machine

Feedback to the machine affecting its operation

Paychecks

Updated master file

Answers

Reports

FIG. 4-2

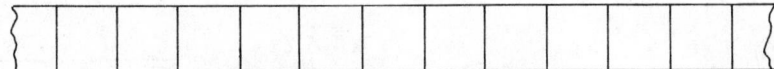

FIG. 4-3 Computer's memory compared to long string of pigeonholes.

user, and, if required, to replace that character in a millionth of a second. Thus a portion of the computer's memory may look as in Fig. 4-4 at one moment and as in Fig. 4-5 an instant later.

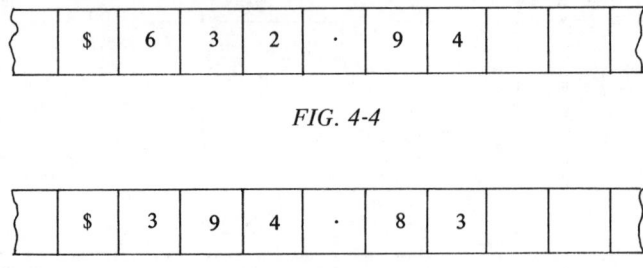

FIG. 4-4

FIG. 4-5

In many computers, each pigeonhole is directly accessible by a computer program. The pigeonholes are called byte locations and a computer is said to have a capacity of 16K bytes, 32K bytes, 64K bytes, etc. A K represents 1024 byte positions; thus a 16K machine has a 16,384-byte memory and a 64K machine has a 65,536-byte memory.

We have used some poetic license in comparing the memory of a computer to a long string of pigeonholes. Actually, a byte location is made up of electronic components. The exact configuration of these components is not important for our purposes at the moment, but it is important to know that each byte location consists of eight almost invisible metallic devices called cores. A core may be magnetized to represent either a zero or a 1. Codes have been developed so that various combinations of zeroes and ones (bits) may represent digits, alphabetic letters, or special signs. Figure 4-6 shows partially one of those codes as used on IBM-360 computer.

The bit representations for the various characters have been shown in the base 16 number system. Thus, if you wish to see what

W looks like, look up W in the table and then convert it to binary. W is $E6_{16}$, which converts directly to

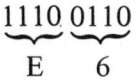

Character	360 Code	Character	360 Code
A	C1	Y	E8
B	C2	Z	E9
C	C3	0	F0
D	C4	1	F1
E	C5	2	F2
F	C6	3	F3
G	C7	4	F4
H	C8	5	F5
I	C9	6	F6
J	D1	7	F7
K	D2	8	F8
L	D3	9	F9
M	D4	/	61
N	D5	&	50
O	D6	blank	40
P	D7	-	60
Q	D8	$	5B
R	D9	.	4B
S	E2	,	6B
T	E3	#	7B
U	E4	*	5C
V	E5	%	6C
W	E6	@	7C
X	E7	'	7D

FIG. 4-6

Since there are 256 different ways that zeroes and ones may be combined in a byte, there are 256 different characters that may be represented in a byte location. These characters may be digits, letters of the alphabet (both upper and lowercase), and many special symbols.

Much of the information that a computer system is to deal with originates on punched cards. A punched card, often called an *IBM card*, is $7\frac{3}{8}$ inches long and $3\frac{3}{4}$ inches wide. It has the capacity to record, in punched form, up to 80 characters of information. Figure

4-7 is an illustration of a punched card. This card shows an employee name, social security number, and hourly pay rate. Note that the employee's last name begins in column 1 of the card, his initials begin in column 11, his social security number begins in column 16, and his pay rate begins in column 28. Observe that every character must occupy a full column of the card (even periods and dashes) and that you may punch only one character per column. (In data processing, periods and decimal points are the same.)

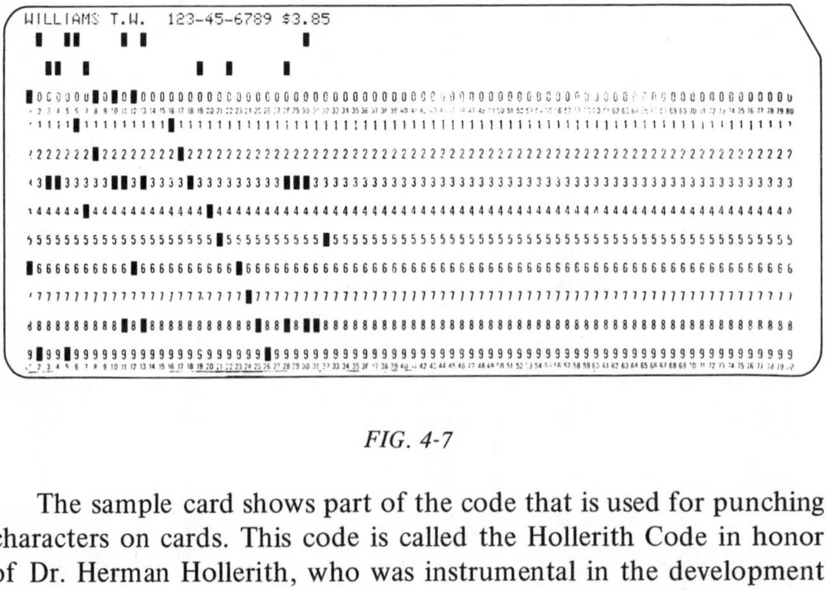

FIG. 4-7

The sample card shows part of the code that is used for punching characters on cards. This code is called the Hollerith Code in honor of Dr. Herman Hollerith, who was instrumental in the development of punched-card data processing.

Study the card and you will see that all 80 columns are numbered. There are 12 rows on the card, also numbered. The bottom 10 rows are numbered 0, 1, 2, 3, etc., while the top two rows are numbered 12 and 11 (reading from top to bottom).

In the Hollerith Code, letters of the alphabet require two punches in a single column; digits require one punch in a single column; and most special characters such as $, -, and . require three punches.

It is not necessary that you memorize the Hollerith Code; however, you may need to refer to Figure 4-8, which gives the Hollerith Code for the characters most used in data processing.

A keypunch machine is used to punch these characters on cards.

Character	Hollerith Code	Character	Hollerith Code
A	12-1	0	0
B	12-2	1	1
C	12-3	2	2
D	12-4	3	3
E	12-5	4	4
F	12-6	5	5
G	12-7	6	6
H	12-8	7	7
I	12-9	8	8
J	11-1	9	9
K	11-2	$	11-3-8
L	11-3	.	12-3-9
M	11-4	(	12-5-8
N	11-5	)	11-5-8
O	11-6	,	0-3-8
P	11-7	space	blank
Q	11-8	%	0-4-8
R	11-9	&	12
S	0-2	+	12-6-8
T	0-3	-	11
U	0-4	*	11-8-4
V	0-5	/	0-1
W	0-6	?	0-7-8
X	0-7	"	5-8
Y	0-8	'	12-7-8
Z	0-9	;	11-6-8
		:	2-8
		@	4-8

FIG. 4-8

The machine has a keyboard much like that of a typewriter. When the operator wishes to punch the letter A, he or she punches the key labeled A. The two holes representing 12 and 1 are simultaneously punched on the card in a single column. The machine then moves the card to the next column position and waits for the operator to strike the next key. When the operator wishes to leave a column blank, he hits the space bar.

The keypunch machine can be used in various semiautomatic ways such as for duplicating or skipping fields, but we shall not go into an explanation of that in this text.

When one or more cards have been punched and the cards are ready to be used by a computer system, they are placed in a computer's *card reader*. A card reader is a device that accepts cards one at a time and passes them through a *read* station. The read station mechanically or optically senses the positions of the holes on the cards and transfers the information thereon to the memory of the computer. Many card readers are fast and are able to read as many as 1000 or 2000 cards per minute.

When a card is read, the information on the card is copied into 80 consecutive byte locations of the computer system. Observe carefully that the Hollerith Code and the internal base 16 code representing individual characters are not the same. For example, the Hollerith Code for the letter A is 12 and 1, but the base 16 code for the letter A is $C1_{16}$. Similarly, the Hollerith Code for the digit 3 is 3, but the base 16 code is $F3_{16}$. Again, the Hollerith Code for a blank is a column with no punched hole, but the base 16 code for a blank is 40_{16}. The conversion from the Hollerith Code to the base 16 code is automatically made by the computer.

Keep in mind that base 16 representations are used for convenience. Actually, every base 16 value shown above becomes an 8-bit binary number in the computer's memory.

In the sample card, the first 32 column positions had information punched in them. This information was stored in 80 consecutive byte positions of memory. In memory, therefore, byte positions corresponding with card positions 33 through 80 have *blank* codes 40_{16}. The first 32 positions of those 80 positions have stored information represented by these codes:

Column	1	2	3	4	5	6	7	8	9	10
Information	W	I	L	L	I	A	M	S		
Base 16 code	E6	C9	D3	D3	C9	C1	D4	E2	40	40

Column	11	12	13	14	15	16	17	18	19	20	21
Information	T	.	W	.		1	2	3	-	4	5
Base 16 code	E3	4B	E6	4B	40	F1	F2	F3	60	F4	F5

Column	22	23	24	25	26	27	28	29	30	31	32
Information	-	6	7	8	9		$	3	.	8	5
Base 16 code	60	F6	F7	F8	F9	40	5B	F3	4B	F8	F5

A programmer knows, either directly or indirectly, where information is in memory. He also knows its content. He does not know details, but he may know, for example, that byte positions 2000_{16} through $200E_{16}$ contain an employee's name, that byte positions $200F_{16}$ through 2019_{16} contain a social security number, etc.

When a programmer wishes to have the computer solve a problem, he gives it a series of instructions. These instructions constitute a *program*. A program may have as few as one or two instructions or as many as several thousand. A programmer initially writes the instructions on a sheet of paper, then has them punched on cards, the same type of card that we have already described, and then feeds the program into a computer's memory. The computer begins to execute the instructions; that is, it does what the programmer wanted him to do.

Computer instructions must be written in a language that the computer understands. That language is flexible and through its use a programmer can get the computer to do many wonderful things. But the language has highly rigid rules which *must* be adhered to.

A computer can be instructed to add two numbers, but it cannot be instructed to fly. It can be instructed to find the smallest of several numeric values, but it cannot be instructed to reveal next week's front-page news.

What are some of the instructions that a programmer may give a computer? Let us examine some examples. It can be commanded to read a punched card containing two values, add them, and then print the result. Suppose the card looks as in Fig. 4-9. We need a command

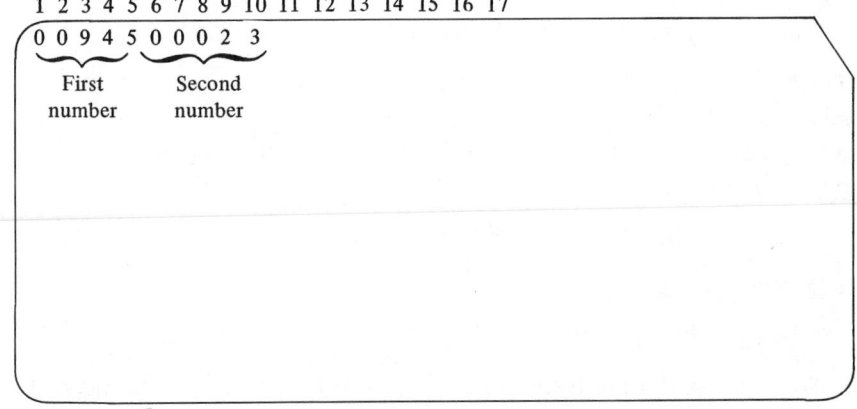

FIG. 4-9

that causes a computer to read the card and bring the information thereon into memory. Here is a command that, while not actually used on any computer that we know of, is typical of a command that does the job we want it to do:

READ DATA CARD AND PLACE CONTENTS
BEGINNING AT BYTE LOCATION 8000.

You can see that the command has a *fixed portion*, which is always written the same way:

READ DATA CARD AND PLACE CONTENTS
BEGINNING AT BYTE LOCATION

and a *variable part*, which tells *where* you wish the information to be stored. In the example, we have arbitrarily selected byte location 8000_{16} as the *beginning* byte location for the 80 characters on the card. Those 80 characters will be stored in byte locations 8000_{16} through $804F_{16}$. As an exercise, verify the following:

These are the base 16 codes representing the first 10 of those byte positions:

Position	8000	8001	8002	8003	8004	8005	8006	8007	8008	8009
Information	0	0	9	4	5	0	0	0	2	3
Base 16 code	F0	F0	F9	F4	F5	F0	F0	F0	F2	F3

The actual 80 bits in memory are these:

11110000111100001111100111110100111110101

11110000111100001111000011110010111110011

These bits are *not* in a form that permits us to have the computer add them. The two numbers must first be converted to 32-bit integer form. This means that the contents of the 5 bytes beginning at byte position 8000_{16} and 8005_{16} must be changed from

$F0_{16}$, $F0_{16}$, $F9_{16}$, $F4_{16}$, $F5_{16}$ to $000003B1_{16}$

and from

$F0_{16}$, $F0_{16}$, $F0_{16}$ $F2_{16}$, $F3_{16}$ to 00000017_{16}

We have said that bytes must be converted to 32-bit integers. It depends on the machine you are using whether the bytes must be converted to 32-bit, 36-bit, 48-bit integers, etc. We have selected 32 bits for these examples to match IBM-360 usage.

The computer you are using probably has an instruction that

causes the conversion of a series of digits in byte format to a 32-bit integer. Here is one version of that command:

CONVERT BYTES 8000-8004 TO INTEGER AND STORE
RESULT AT WORD LOCATION 9000.

The instruction has two fixed portions:

CONVERT BYTES

and

TO INTEGER AND STORE RESULT AT WORD LOCATION

It also has two variable portions, the first telling where the digits to be converted are stored and the second telling where to place the result. Since the five digits beginning at byte location 8000_{16} are converted to a complete 32-bit word, the word requires four byte locations in memory. Therefore the 32-bit result will be stored at byte locations 9000_{16} through 9003_{16}.

We use the same instruction to convert the five digits located at byte positions 8005_{16} through 8009_{16}:

CONVERT BYTES 8005-8009 TO INTEGER AND STORE
RESULT AT WORD LOCATION 9004.

It is of interest to note that 32 bits may express an integer having a value in excess of 2 billion. Therefore, when using the above instruction, you may convert a number containing up to ten sequential digits.

We must now give the command that will cause the two integers located at word location 9000_{16} and 9004_{16} to be added:

ADD INTEGER 9000 AND 9004 GIVING 9008.

The instruction works on complete 32-bit words; therefore the instruction simply mentions the first byte location of each word. The result is given as an integer.

To print an answer, in this example the contents of the word beginning at byte location 9008_{16}, the word must be converted to 8-bit bytes. This instruction will accomplish the task:

CONVERT INTEGER 9008 TO BYTES AND STORE
RESULT AT 7000-7005.

We know that the answer is 968 (00945 + 00023) but we have given 6 byte positions as the storage area. The result will be stored there as 000968. In memory, the result will be stored like this:

Position	7000	7001	7002	7003	7004	7005
Information	0	0	0	9	6	8
Base 16 code	F0	F0	F0	F9	F6	F8

The programmer has to have some idea about the number of digits expected in the answer; otherwise he would not know how many byte positions to allow for the result.

When characters are stored in byte format, they may be printed using a PRINT instruction:

PRINT 7000-7005 BEGINNING AT PRINTING POSITION 15.

This instruction prints on paper the character contents of byte positions 7000_{16} through 7005_{16}. It begins printing on print position 15 of the paper. (In most computers, at least 120 print positions are available.) On the answer paper the value 000968 will be printed beginning at print position 15.

The PRINT instruction may include an optional feature to suppress leading zeroes. If so, the instruction is written like this:

PRINT 7000-7005 BEGINNING AT PRINT POSITION 15
SUPPRESSING LEADING ZEROES.

The effect of this instruction is to change leading zeroes to blanks. The characters that are printed will be blank, blank, blank, 9, 6, 8.

The entire list of instructions constituting the program to solve the problem are

READ DATA CARD AND PLACE CONTENTS BEGINNING AT BYTE LOCATION 8000.

CONVERT BYTES 8000-8004 TO INTEGER AND STORE RESULT AT WORD LOCATION 9000.

CONVERT BYTES 8005-8009 TO INTEGER AND STORE RESULT AT WORD LOCATION 9004.

ADD INTEGER 9000 AND 9004 GIVING 9008.

CONVERT INTEGER 9008 TO BYTES AND STORE
RESULT AT 7000-7005.

PRINT 7000-7005 BEGINNING AT PRINT POSITION 15
SUPPRESSING LEADING ZEROES.

STOP RUN.

The STOP RUN instruction tells the computer that the job is done. Here is a summary of the instructions we have learned so far and a few new ones:

READ DATA CARD AND PLACE CONTENTS BEGINNING
AT BYTE LOCATION byte location.

CONVERT BYTES byte locations TO INTEGER AND
STORE RESULT AT WORD LOCATION word location.

CONVERT INTEGER word location TO BYTES AND
STORE RESULT AT byte locations.

ADD INTEGER word location AND word location GIVING
word location.

PRINT byte locations BEGINNING AT PRINT POSITION
print position.

PRINT byte locations BEGINNING AT PRINT POSITION
print position SUPPRESSING LEADING ZEROES.

STOP RUN.

CONVERT BYTES byte locations TO FLOATING POINT
AND STORE RESULT AT WORD LOCATION word location.

MULTIPLY FLOATING POINT word location BY word
location GIVING word location.

CONVERT FLOATING POINT word location TO BYTES
AND STORE RESULT AT byte locations.

Note that the fixed parts of the instructions are shown in capital letters. You must copy these words exactly as they are shown. Values that you are expected to fill in are shown in lowercase letters. Thus in the last instruction of the list, you must *copy* the words CONVERT FLOATING POINT and TO BYTES AND STORE

RESULT AT, and you must supply a word location and a range of byte locations where the lowercase words *word location* and *byte locations* are shown.

Here is another problem: Assume that the data card in Fig. 4-10 must be processed. The problem is to multiply the first value by the second value and print the result.

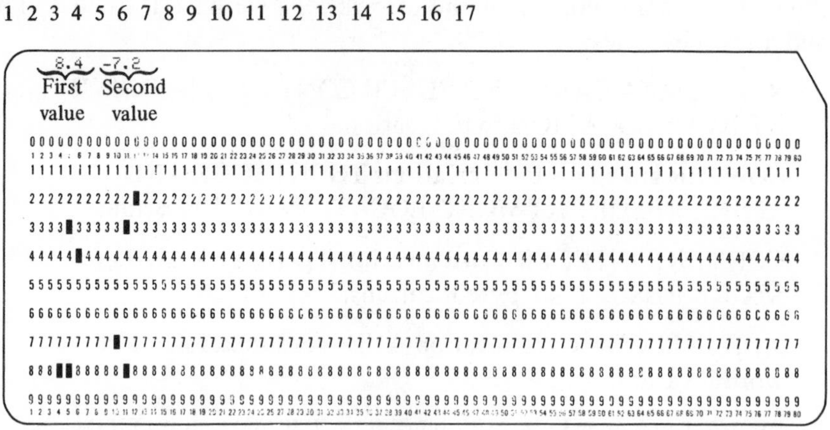
1 2 3 4 5 6 7 8 9 10 11 12 13 14 15 16 17
8.4
-7.2
First value
Second value

FIG. 4-10

Here is the program that does the job:

READ DATA CARD AND PLACE CONTENTS BEGINNING AT BYTE LOCATION 2000.

CONVERT BYTES 2000-2005 TO FLOATING POINT AND STORE RESULT AT WORD LOCATION 3000.

CONVERT BYTES 2006-200B TO FLOATING POINT AND STORE RESULT AT WORD LOCATION 3004.

MULTIPLY FLOATING POINT 3000 BY 3004 GIVING 3008.

CONVERT FLOATING POINT 3008 TO BYTES AND STORE RESULT AT 4000-4007.

PRINT 4000-4007 BEGINNING AT PRINT POSITION 20 SUPPRESSING LEADING ZEROES.

STOP RUN.

The program is similar to the one shown earlier except that the numbers include decimal points. We cannot, therefore, employ integer arithmetic but must call for floating-point arithmetic.

Problem: Expressed as base 16 numbers, what are the digits stored at word locations 3000_{16}, 3004_{16}, and 3008_{16}? Answer: $41866666_{16} = 8.4$; $C1733333_{16} = -7.2$; $C23C7AE1_{16}$ (-60.48).

The answer will be printed beginning on column 20 of the printed paper. The eight characters will be blank, blank, minus sign, 6, zero, decimal point, 4, 8.

Consider another problem where gross wages are to be calculated from information given on data cards. Each card will have a format as shown in Fig. 4-11. The first 11 columns contain the social security number of the person whose gross wage is being computed; the next 6 columns contain the hours he worked last week; and the next 5 columns contain his pay rate per hour.

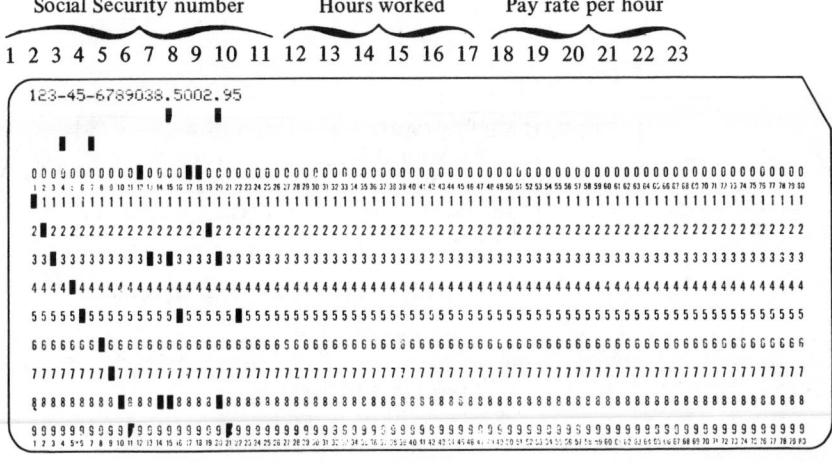

FIG. 4-11

There are an indeterminate number of cards in the data deck, but the last one has the format shown in Fig. 4-12. The problem is to have a program read a data card for each person, compute gross pay, and print gross pay and social security number. The program is to stop when the last card, having the special format of Fig. 4-12 is detected.

The usual first step in writing a program is to develop a flowchart.

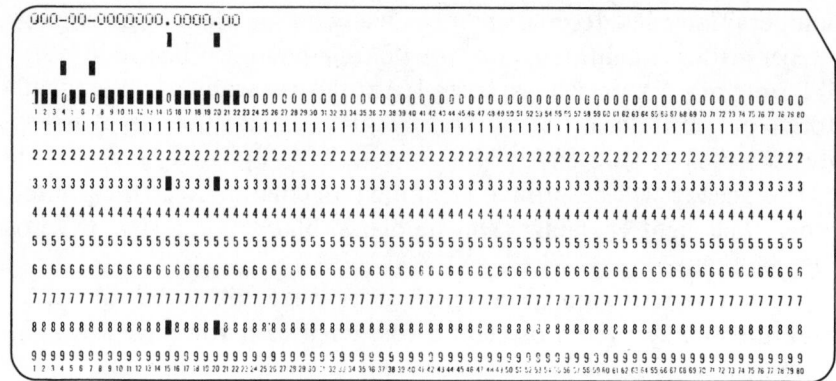

FIG. 4-12

The flowchart that would correspond to a solution of the problem is shown in Fig. 4-13.

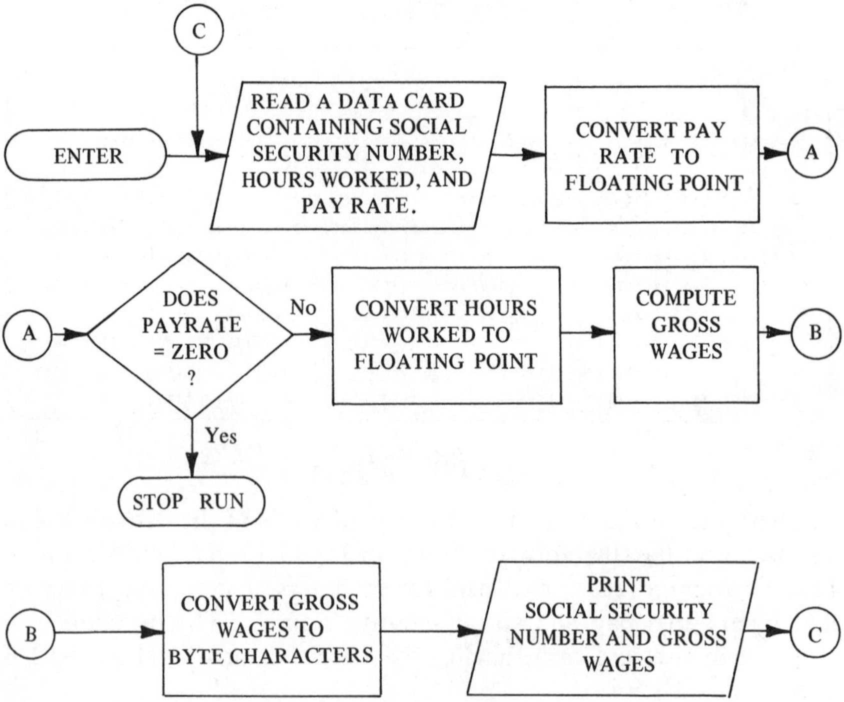

FIG. 4-13

A flowchart is a pictorial plan showing what you want the computer to do and in what sequence. It is meant to be used as a guide when a programmer is writing a program and to serve as partial documentation when the program has been written.

Most problems are much more complex than the one given here and a flowchart may extend over dozens of pages.

You can see that various shaped boxes are used to indicate various tasks to be done. A rectangle indicates some processing such as converting a number from byte format to floating-point or multiplying one floating-point number by another.

A diamond indicates a decision to be made by the program. There must be at least two paths flowing from a diamond.

The parallelogram is used for input-output (I/O) operations, i.e., transferring data from input to storage, or from storage to an output medium.

Arrows show the path that the processing must take. A person begins with the oval labeled ENTER and then follows the arrows. Notice how connectors (small circles) are used to connect one part of a flowchart with another.

Let us solve the problem given above using the instructions we already know and some new ones.

BEGIN.   READ DATA CARD AND PLACE CONTENTS BEGINNING AT BYTE LOCATION 1000.

CONVERT BYTES 1011-1015 to FLOATING POINT AND STORE RESULT AT WORD LOCATION 2000.

DOES CONTENTS OF WORD 2000 EQUAL ZERO; IF YES GO TO TERMINATE; IF NO GO TO NEXT INSTRUCTION.

CONVERT BYTES 100B-1010 TO FLOATING POINT AND STORE RESULT AT WORD LOCATION 2004.

MULTIPY 2000 BY 2004 GIVING 2008.

CONVERT FLOATING POINT 2008 TO BYTES AND STORE RESULT AT 2012-202B.

PRINT 1000-100A BEGINNING AT PRINT
POSITION 10; PRINT 2012-202B
BEGINNING AT PRINT POSITION 30
SUPRESSING LEADING ZEROES.

GO TO BEGIN.

TERMINATE. STOP RUN.

As a programmer writes instructions, it would probably be a good idea for him to make notes he can use for reference. For example, the data card will be read into the 80 byte locations of memory beginning with location 1000_{16}. The programmer observes that the social security number will be located at byte position 1000_{16} through $100A_{16}$, that the hours worked will be located at byte position $100B_{16}$ through 1010_{16}, and that the pay rate will be located at byte positions 1011_{16} through 1015_{16}.

Later in the program, the programmer causes the program to make certain conversions and to store the results at arbitrarily selected locations. As the programmer works, he may jot down these notes on a sheet of paper:

Item C	Byte Locations and Size		Comments
Social security number (bytes)	1000-100A	11	From card
Hours worked (bytes)	100B-1010	6	From card
Pay rate (bytes)	1011-1015	5	From card
Pay rate (floating point)	2000-2003	4	Converted by program
Hours worked (floating point)	2004-2007	4	Converted by program
Gross wages (floating point)	2008-2011	4	Computed by program
Gross wages (bytes)	2012-202B	10	Converted by program

There are several new ideas in the coded program. First note the *labels* BEGIN and TERMINATE. Labels are used to identify various parts of a program. They may be any descriptive names you select, not necessarily BEGIN and TERMINATE.

Note that the first instruction of the program needs a label because of the GO TO instruction near the end. Also, the last instruction of the program needs a label because of the conditional jump in the instruction that begins with the word DOES.

We may assume that the word ZERO (see third instruction) is understandable to the computer and represents either a floating-point or fixed-point zero (32 zero bits in a word).

Note that the PRINT instruction prints one line, not two, because of the semicolon ahead of the second PRINT. If the semicolon had been a period, the program would have printed two lines.

On each line of the printed output, this program prints a social security number and the gross pay for the person involved. The gross pay that was computed and stored as a floating-point number must be converted to byte form before it can be printed. The social security number, on the other hand, needs no conversion because it was read from a punched card in byte form and must be used in that form for output. (As you will recall, the PRINT instruction demands that characters be in 8-bit byte format.)

You should bear in mind that the programming language presented here does not exist. Nevertheless, it is typical of programming languages that do exist. Actual languages are sometimes more brief, sometimes more wordy. For example, the instruction

DOES CONTENTS OF WORD 2000 EQUAL ZERO; IF YES
GO TO TERMINATE; IF NO GO TO NEXT INSTRUCTION.

could be written more briefly as

IF WORD 2000 = ZERO GO TO TERMINATE; ELSE NEXT
INSTRUCTION.

The conditional instruction given above is self-explanatory. A condition is tested (in the example, the condition is WORD 2000 = ZERO). If the condition is true, the program is directed to take the jump indicated (GO TO TERMINATE); if the condition is false, the program is directed to take the alternative path (ELSE NEXT INSTRUCTION).

The terseness or wordiness of a language is determined by the people who invent it and offer it to programmers. Every computer must have at least one language by which it may be programmed, and the specifications of that language are written in a manual and delivered with the computer. This language is called the computer's *machine language* and, in general, the language is not acceptable for programming other computers.

Most modern computers also make available one or more

universal language. A programmer who knows a universal language has the ability to write programs for a wide variety of computers, not necessarily all the same make.

The most widely used universal languages are FORTRAN, COBOL and PL/I. FORTRAN is a scientific programming language, COBOL is a business language, and PL/I a combination scientific/ business language.

There are books and manuals available for students who wish to learn how to write programs in machine or universal languages. It is not within the scope of this text to delve too deeply into how to use these languages, but we can give you examples of programs written in machine and universal languages.

The machine language delivered with the GE-635 computer, as an example, is called GMAP. Figure 4-14 shows a program written in GMAP.

Without going into detail regarding what each individual instruction does, we shall simply say that this program merges two or more business files and sorts them in a predetermined sequence.

Here is a program written in FORTRAN:

```
        SUM = 0.
        KNT = 0
  8     READ 3,X
        IF (X.EQ.999.) GØ TØ 25
        KNT = KNT + 1
        SUM = SUM + X
        GØ TØ 8
 25     AVG = SUM/KNT
        PRINT 4, SUM,AVG
        STØP
  3     FORMAT (F10.0)
  4     FORMAT (2F15.2)
        END
```

```
          600SH
)         SORT      INOUT
          FIELD     (W1,C6,C24)
          SEQ       (A3,A2)
)         UCE       ICE
          FILCB     INOUT,**,2,,,1,26
    ICE   EQU       *
)         CALL      OPEN(IFD,2)
          CALL      GET(ISFCB,EOFI)
    PI    LDA       TAL,SC
)         TTF       *+2
          TRA       EOFI
          CMPA      =020,DL
)         TZE       PI
          EAXO      ALLDON
          STXO      *-2
)         CMPA      =020,DL
          TRC       EOFI
          STA       TALO,SC
)         TTF       PI
    EOFI  LDQ       =0H0SX,DL
          MME       GEBORT
)   ALLDON LDQ      TALO
          ANQ       =077,DL
          TZF       EOFI
)         NEGL
          ADQ       3,DL
          MPY       6,DL
)         ADQ       54,DL
          EAXO      0,QL
          LDA       TEMP
)         LRI       0,0
          CALL      .GPNRY
          NEGL
)         STO       TEMP
    S1    CALL      GET(INFILE,FOF)
          LDX1      INFILE
)         STX1      ARG2
    S2    CALL      .SPUT(0,**)
          TRA       S1
)   ARG2  EQU       S2+4
    EOF   CALL      CLOSE(IFD,2)
          CALL      .SPUTC
)   IFD   VFD       18/INFILE,1/0,1/1,2/3
          VFD       18/ISFCB,1/0,1/1,2/3
          FILCB     INFILE,IN,BUF26,BUF26A,,1,26,,,,,,,,,,EOR
)         F&LCB     ISFCB,IS,IBUFF
    EOR   AOS       TEMP
          TZE       0,1
)         LDXO      3,1
          LDQ       EORI
          STO       0,0
)         TRA       0,1
    EOR1  BCI       1,EOR
    IBUFF BSS       321
)
    TAL   TALLY     IBUFF+3,81,0
    RALO  TALLY     TEMP,4,0
)   BUF26 BSS       321
    TEMP  BSS       1
    BUF26A BSS      321
)         END
```

FIG. 4-14

A popular programming language available to business programmers is COBOL. The letters in the word COBOL stand for *CO*mmon *B*usiness *O*riented *L*anguage. This language is ideally suited for working with files. It uses English words, which enables programmers to learn it quickly and employ it efficiently. The program in Fig. 4-15 is wirtten in COBOL. This program updates a business file. As you can see, programs written in COBOL are English-oriented. COBOL has a special vocabulary that enables people to instruct a computer using words in that vocabulary. In addition to the built-in vocabulary, a programmer may invent words of his own to express certain ideas. The only requirement is that he explain the meanings of those words in the appropriate places of his program.

COBOL PROGRAM SHEET

PROGRAM		COMPUTER		CHAR. SET		JOB #		DATE		SHEET		OF
PROGRAMMER		BLOG.	RM	PICK UP ☐	CALL EXT.		DELIVER ☐	DATE REQ'D		SOURCE 73 DECK IDENT. 80		
SEQUENCE NUMBER	C	PRINT ON ☐	SEQ. DECK ☐	INTERPRET ☐	LIST ☐	SW ☐	ON ☐	PAPER SIZE	PARTS			

```
100100  IDENTIFICATION DIVISION.
100110  PROGRAM-ID. UPDATE.
200100  ENVIRONMENT DIVISION.
200110  CONFIGURATION SECTION.
200120  SOURCE-COMPUTER. GE-635.
200130  OBJECT-COMPUTER. GE-635.
200140  INPUT-OUTPUT SECTION.
200150  FILE-CONTROL. SELECT MASTER-FILE ASSIGN TO M1. SELECT
200160         TRANSACTION-FILE ASSIGN TO T1. SELECT NEW-MASTER-FILE ASSIGN
200170         TO N1. SELECT REPORT-FILE ASSIGN TO R1 FOR LISTING. SELECT
200180         SORT-FILE ASSIGN TO Z1, Z2, Z3.
300100  DATA DIVISION.
300110  FILE SECTION.
300120  FD  MASTER-FILE, BLOCK CONTAINS 100 RECORDS; LABEL RECORDS ARE
300130         STANDARD, VALUE OF ID IS "INVENTORY"; DATA RECORD IS
300140         MASTER-RECORD.
300160  01  MASTER-RECORD.
300170      02  PART-NUM          PICTURE 9(5).
300180      02  DESCRIPTION       PICTURE X(15).
300190      02  VENDOR            PICTURE X(25).
300200  FD  TRANSACTION-FILE, BLOCK CONTAINS 100 RECORDS; LABEL RECORDS
300210         ARE STANDARD, VALUE OF ID IS "TRANSACTIONS"; DATA RECORD IS
300220         TRANSACTION-RECORD.
300240  01  TRANSACTION-RECORD.
300250      02  PART-NUM          PICTURE 9(5).
300260      02  DESCRIPTION       PICTURE X(15).
300270      02  VENDOR            PICTURE X(25).
300280      02  KODE              PICTURE 9.
300290      88  CHANGE VALUE IS 1.
300300  FD  NEW-MASTER-FILE; BLOCK CONTAINS 100 RECORDS; LABEL RECORDS
300310         ARE STANDARD; VALUE OF ID IS "INVENTORY"; RETENTION-PERIOD
300320         020; DATA RECORD IS NEW-MASTER-RECORD.
300340  01  NEW-MASTER-RECORD.
300350      02  PART-NUM          PICTURE 9(5).
300360      02  DESCRIPTION       PICTURE X(15).
```

FIG. 4-15

3,0,0,3,7,0	`02 VENDØR PICTURE X(25).`
3,0,0,3,9,0	`FD REPØRT-FILE; LABEL RECØRDS ARE STANDARD.; REPØRT IS`
3,0,0,4,0,0	`ADDITIØN-REPØRT.`
3,0,0,4,1,0	`SD SØRT-FILE; DATA RECØRD IS SØRT-RECØRD.`
3,0,0,4,2,0	`01 SØRT-RECØRD.`
3,0,0,4,3,0	`02 PART-NUM PICTURE 9(5).`
3,0,0,4,4,0	`02 DESCRIPTIØN PICTURE X(15).`
3,0,0,4,5,0	`02 VENDØR PICTURE X(25).`
3,0,0,4,6,0	`02 KØDE PICTURE 9.`
3,0,0,4,7,0	`88 CHANGE VALUE IS 1.`
3,0,0,4,8,0	`WØRKING-STØRAGE SECTIØN.`
3,0,0,4,9,0	`01 WS-RECØRD.`
3,0,0,5,0,0	`02 P-NUM PICTURE 9(5).`
3,0,0,5,1,0	`02 DESCRIPT PICTURE X(15).`
3,0,0,5,2,0	`02 VENDØR-INFØ PICTURE X(25).`
3,0,0,5,3,0	`REPØRT SECTIØN.`
3,0,0,5,4,0	`RD ADDITIØN-REPØRT; PAGE LIMITS ARE 50 LINES.; HEADING 3.`
3,0,0,5,5,0	`01 TYPE PAGE HEADING.; NEXT GRØUP PLUS 2.`
3,0,0,5,6,0	`02 LINE-ØNE.; LINE NUMBER PLUS 1.`
3,0,0,5,7,0	`03 CØLUMN NUMBER 58.; PICTURE A(15).; VALUE IS`
3,0,0,5,8,0	`"ADDITIØN REPØRT".`
3,0,0,5,9,0	`02 LINE-TWØ.; LINE NUMBER PLUS 2.`
3,0,0,6,0,0	`03 CØLUMN NUMBER 5.; PICTURE A(11).; VALUE IS`
3,0,0,6,1,0	`"PART NUMBER".`
3,0,0,6,2,0	`03 CØLUMN NUMBER 20.; PICTURE A(11).; VALUE IS`
3,0,0,6,3,0	`"DESCRIPTIØN".`
3,0,0,6,4,0	`03 CØLUMN NUMBER 45.; PICTURE A(18).; VALUE IS`
3,0,0,6,5,0	`"VENDØR AND ADDRESS".`
3,0,0,6,8,0	`01 DETAIL-LINE TYPE DETAIL.; LINE NUMBER PLUS 1.; NEXT GRØUP`
3,0,0,6,9,0	`PLUS 1.`
3,0,0,7,0,0	`02 CØLUMN NUMBER 10.; PICTURE 9(5).; SØURCE IS P-NUM.`
3,0,0,7,1,0	`02 CØLUMN NUMBER 20.; PICTURE X(15).; SØURCE IS DESCRIPT.`
3,0,0,7,2,0	`02 CØLUMN NUMBER 45.; PICTURE X(25).; SØURCE IS VENDØR-INFØ.`
4,0,0,1,0,0	`PRØCEDURE DIVISIØN.`
4,0,0,1,0,5	`PRØCESS-SØRT SECTIØN.`
4,0,0,1,1,0	`SØRT-PRØCEDURE. SØRT SØRT-FILE ØN ASCENDING KEY PART-NUM ØF,`
4,0,0,1,2,0	`SØRT-RECØRD USING TRANSACTIØN-FILE ØUTPUT PRØCEDURE REC-PRØC.`
4,0,0,1,3,0	`STØP RUN.`
4,0,0,1,5,0	`REC-PRØC SECTIØN.`
4,0,0,1,6,0	`ØPEN-PRØC. ØPEN INPUT MASTER-FILE ØUTPUT NEW-MASTER-FILE.;`
4,0,0,1,7,0	`REPØRT-FILE. INITIATE ADDITIØN-REPØRT.`
4,0,0,1,9,0	`READ-MF-RECØRD. READ MASTER-FILE; AT END GØ TØ TERM-TF-RECØRDS.`
4,0,0,2,1,0	`READ-TF-RECØRD. RETURN SØRT-FILE; AT END GØ TØ TERM-MF-RECØRDS.`
4,0,0,2,3,0	`TEST-PART-NUMS. IF PART-NUM ØF MASTER-RECØRD = PART-NUM ØF`
4,0,0,2,4,0	`SØRT-RECØRD GØ TØ PART-DRØP-TEST.; ØTHERWISE GØ TØ`
4,0,0,2,5,0	`CØMPARE-PART-NUMS.`
4,0,0,2,7,0	`PART-DRØP-TEST. IF CHANGE IN SØRT-RECØRD GØ TØ`
4,0,0,2,8,0	`WRITE-TRANS-RECØRD.; ØTHERWISE GØ TØ READ-MF-RECØRD.`
4,0,0,3,0,0	`WRITE-TRANS-RECØRD. MØVE SØRT-RECØRD TØ WS-RECØRD.; WRITE`
4,0,0,3,1,0	`NEW-MASTER-RECØRD FRØM WS-RECØRD. GØ TØ READ-MF-RECØRD.`
4,0,0,3,3,0	`CØMPARE-PART-NUMS. IF PART-NUM ØF MASTER-RECØRD < PART-NUM ØF,`
4,0,0,3,4,0	`SØRT-RECØRD GØ TØ WRITE-MF-RECØRD.; ØTHERWISE NEXT SENTENCE.`
4,0,0,3,6,0	`WRITE-TF-RECØRD. MØVE SØRT-RECØRD TØ WS-RECØRD. WRITE`
4,0,0,3,7,0	`NEW-MASTER-RECØRD FRØM WS-RECØRD.; GENERATE DETAIL-LINE.; GØ`
4,0,0,3,8,0	`TØ READ-TF-RECØRD.`
4,0,0,4,0,0	`WRITE-MF-RECØRD. WRITE NEW-MASTER-RECØRD FRØM MASTER-RECØRD.. READ`
4,0,0,4,1,0	`MASTER-FILE; AT END GØ TØ TERM-TF-RECØRDS.. GØ TØ`
4,0,0,4,2,0	`TEST-PART-NUMS.`
4,0,0,4,4,0	`TERM-TF-RECØRDS. MØVE SØRT-RECØRD TØ WS-RECØRD.. WRITE`
4,0,0,4,5,0	`NEW-MASTER-RECØRD FRØM WS-RECØRD.. GENERATE DETAIL-LINE..`
4,0,0,4,6,0	`RETURN SØRT-FILE.; AT END GØ TØ CLØSE-ØUT.. GØ TØ`
4,0,0,4,7,0	`TERM-TF-RECØRDS.`
4,0,0,4,9,0	`TERM-MF-RECØRDS. WRITE NEW-MASTER-RECØRD FRØM MASTER-RECØRD.. READ`
4,0,0,5,0,0	`MASTER-FILE; AT END GØ TØ CLØSE-ØUT.. GØ TØ TERM-MF-RECØRDS.`
4,0,0,5,2,0	`CLØSE-ØUT. CLØSE MASTER-FILE, NEW-MASTER-FILE.. TERMINATE`
4,0,0,5,3,0	`ADDITIØN-REPØRT.. CLØSE REPØRT-FILE..`
4,0,0,5,5,0	`EXIT-PØINT. EXIT..`
4,0,0,5,6,0	`END PRØGRAM.`

FIG. 4-15 (cont.)

You can see that a person who does not understand programming would be able to make more sense of a program written in COBOL than one written in GMAP or any of the other so-called machine languages. Understandability was one of the goals desired by the people who invented COBOL.

Now we can discuss the term *compilation*. A person who writes a program in COBOL, FORTRAN, PL/I, or some other universal language similar to these, originally writes the instructions on a coding form. The form is submitted to a cardpunch operator who punches $7\frac{3}{8} \times 3\frac{1}{4}$ inch data processing (IBM) cards from the information on the coding sheet. The operator uses a device called a *key-punch*, which looks very much like an electric typewriter. The operator punches one card for every line on the coding sheet.

A deck of cards is generated as the card-punch operator transforms the information on the COBOL coding form to cards. This deck is called the *source deck*. A programmer takes the source deck to a computer and indicates his intention to obtain a *COBOL compilation*. That is, he says that he wants the instructions on his source deck translated to computer language.

The computer console operator obtains from a storage rack a reel of magnetic tape labeled "COBOL Compiler." This reel contains a computer program that actually does the translating. The computer console operator reads the program from the tape into the memory of the computer. The operator then places the source deck in the computer's card reader and activates the compiler program.

You know, of course, that a computer does not understand English. Suppose that you write this COBOL instruction:

READ PERSONNEL-MASTER-FILE INTO RECORD-PROCESS-AREA; AT END GO TO WINDUP-ROUTINE.

The computer will not be able to understand this statement any better than if it were written in Greek.

However, a computer is able to understand instructions if they are written in a language it does understand. That language involves the use of only two characters, the digits zero and 1.

If the English message above could be translated into meaningful patterns of zeroes and ones, the computer would be able to perform the work you require of it. This is where the COBOL compiler enters the picture. The single function of the COBOL compiler is to transform the English words used in COBOL statements to computer language. Once this has been done, the computer will be able to execute those statements and give the required output.

In a simplified sense, then, a computer acts as its own language translator. It looks up English words in a "dictionary" and substitutes computer-language words made up of zeroes and ones.

Figure 4-16 shows the sequence of events in transforming a COBOL program to computer language. The COBOL compiler

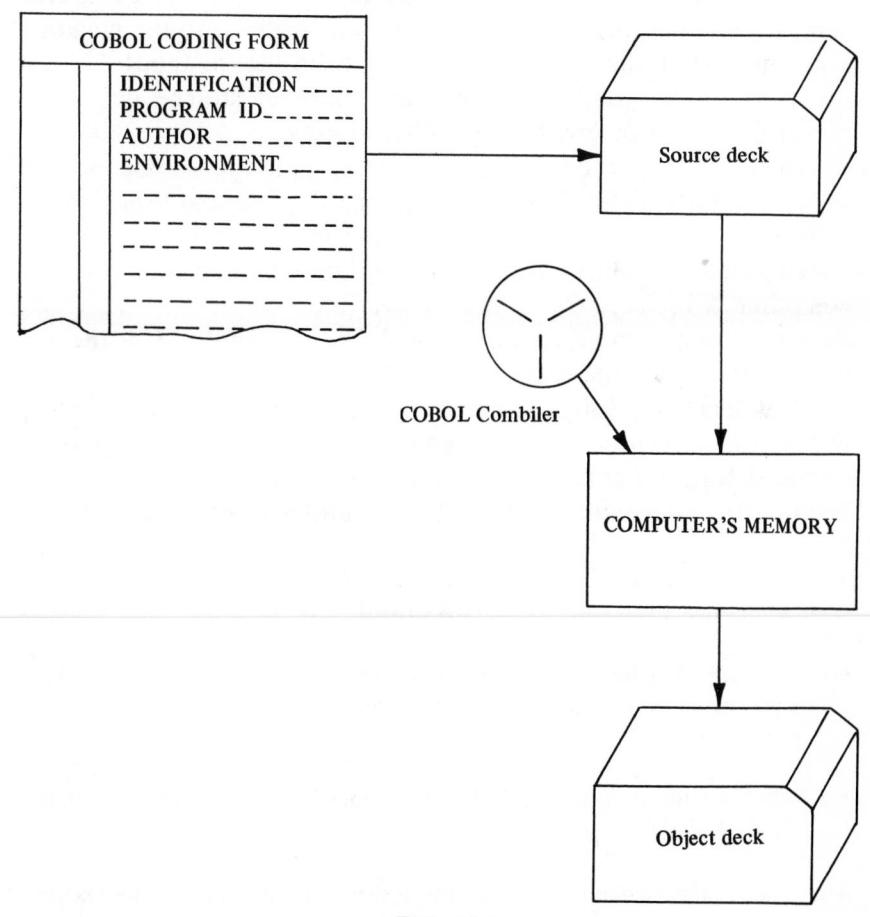

FIG. 4-16

program accepts from the card reader one source card at a time and "decodes" it. During this process, the program determines what machine operations are required as substitutions for the English statements. The correct operations are selected and punched out, by the computer, on a card or series of cards. This procedure is repeated for the second and subsequent source cards.

The deck of cards punched out by the computer is called the *object deck*. The object deck contains the original COBOL instructions but in computer language.

Since the COBOL compiler is no longer required in the memory of the computer, it is erased from memory and the computer-language instructions in the object deck are loaded into the memory. The solution of the original problem is then begun. Input data for the program are obtained from such external storage devices as punched cards, magnetic tape, or other sources.

Once an object deck has been generated, the source deck is stored away. It will not be required, for the present at least, since it is the object deck that actually is used to run programs.

In general, an object deck can be used on only one computer. If a computer is replaced by a more powerful one, the original source deck is retrieved and recompiled for the new computer. It is the *new* object deck that is thereafter used on the new computer.

A programmer solves problems by instructing a computer to help him solve these problems. The ability to solve problems cannot be obtained from a book; however, books such as this one can assist in helping to understand the *ideas* behind problem-solving techniques.

EXERCISES

4-1. What are the IBM-360 hexadecimal codes for the characters P, S, 4, and $? Give the codes that show how these characters are stored in byte positions of memory.

4-2. What are the punched card Hollerith codes for the following characters: H, T, 3, and *.

4-3. What is the beginning word of the instruction described in the text that causes a program to read a punched card?

4-4. Discuss why programming instruction formats require fixed and variable portions.

4-5. Using the instructions described in the text, write a program that causes a data card to be read. The data card contains two five-digit fixed-point values. Have the program multiply one value by the other and print the result.

4-6. What is COBOL?

4-7. Define the term *compilation*

4-8. What is a source deck? Object deck?

Chapter 5

ARRAYS

In data processing, the use of arrays is common. Therefore it is important that you understand what is meant by the term and also how arrays are used.

An array is an arrangement of related numeric or alphanumeric values. For example, study the arrangement (array) in Fig. 5-1. The

	0	1
0	0	0
1	0	1

FIG. 5-1

array represents the binary multiplication table. Using it, you can verify that $0_2 \times 0_2 = 0_2$, $0_2 \times 1_2 = 1_2$, $1_2 \times 0_2 = 1_2$, and $1_2 \times 1_2 = 1_2$. Having spent considerable time working with binary numbers, you should have no difficulty in preparing an array representing the binary addition table (Fig. 5-2).

	0	1
0	0	1
1	1	10

FIG. 5-2

Arrays have dimensions. One-dimensional arrays are called *lists*; two-dimensional arrays are called *tables*. It is also possible to have arrays with more than two dimensions. There is no special name given to three-dimensional, four-dimensional, or other multidimensional arrays.

First, let us discuss lists (one-dimensional arrays). A list is a group of related numeric or alphanumeric values arranged in one-dimensional sequence. Where computers are concerned, these values are generally in consecutive memory words. Consider the example in Fig. 5-3. There are eight cells in the list shown; each cell contains a numeric value.

9.25	6.74	.26	.89	1.63	2.95	8.45	4.50

FIG. 5-3

One of the properties of a list is *size*. The size of the list in Fig. 5-3 is 8 because there are eight cells making up the list and eight values stored in the list. As you can see, one determines the size of a list by counting the number of memory cells assigned to it.

In this discussion, it is not particularly important to concern ourselves with the maximum number of characters that may be found inside a single cell of a list. In general, if the contents of a cell are numeric, the values may be large, such as 3945837.6, or small, such as .00006728, either positive or negative. If the contents of a cell is alphanumeric, the characters may be limited to four, six, eight, etc., depending on which computer is involved. As an example, suppose that a single cell contains a catalog number. That number could be X3-452 or P9-48T, etc.

Each cell of a list has a unique *address*. For example, consider the list we showed above, repeated here as Fig. 5-4. The cell locations are numbered 1 through 8 beginning at the left. Thus the value 9.25 is located at list location 1; the value 6.74 is located at list location 2; etc.

9.25	6.74	.26	.89	1.63	2.95	8.45	4.50

FIG. 5-4

If we give the list a name, such as Q, then we can employ another way to show the contents of the list locations. We may write

$$Q_1 = 9.25$$

$$Q_2 = 6.74$$

$$Q_3 = .26$$

$$\vdots$$

The digits shown to the right and slightly below the name of the list are called *subscripts*. A subscript, therefore, is a number that points to one of the cells of a list.

Before going further, let us review what we have learned about lists. A list is a series of related values (numeric or alphanumeric). In computers, values of a list are placed in sequential memory cells.

Within reason, the size of a list may be anything you wish. (What is reasonable depends on the memory size of the computer you are dealing with. It is not unusual to use lists that comprise several thousand memory cells.)

Lists are named. Often the name of the list will indicate the nature of its contents. For example, names such as CATNUM, SS-NUM, etc., may be used.

You designate any specific list location by using subscripts. For example, if you wish to refer to the sixth location of the CATNUM list, you would refer to it as $CATNUM_6$.

Should the nature of the problem being solved require it, any reasonable number of lists may be used in a single program.

Now let us discuss some of the things that we may want to do with a list once it has been established.

Assume that the list that has been established, R, has the initial values shown in Fig. 5-5. One of the tasks we might want a computer to do is find which value in list R is smallest. A program could be written to have the computer examine every value in the list and print the smallest value, i.e., 3.

R

31	38	15	3	11	3	21	41	9	22

FIG. 5-5

Another task might be to have the computer report that the largest value in list R is 41. The two tasks could be combined so that the computer reports that the smallest value is 3 and the largest value is 41.

We might want the computer to find the *position* of the largest value. The largest value is 41 and is found in cell 8. That location position may be printed.

Suppose that we want the computer to sum the ten values in list R. A program could be written that causes the computer to compute the sum of the ten values and print it. The sum, of course, is 195.

If the average of the ten values is desired, it requires but a single additional step to have the summing program also provide the average, 19.5.

Keep in mind as we discuss these points that the size of a list may be much larger than 10. A list may contain thousands of values. While it is easy for us to compute manually the sum and average of ten numbers, it might not be so easy to accomplish the same task where several thousand values are involved. A computer program would be a definite aid where many values are involved.

Another problem that might arise is that of ordering the values in a list in some definite sequence, such as increasing or decreasing. The ten values in list R are 31, 38, 15, 4, 11, 3, 21, 41, 9, and 22. We might want them rearranged so that list R looks as in Fig. 5-6 or as in Fig. 5-7.

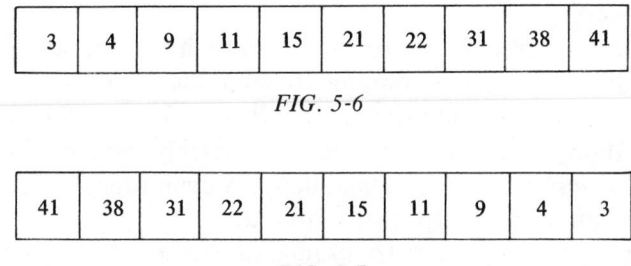

| 3 | 4 | 9 | 11 | 15 | 21 | 22 | 31 | 38 | 41 |

FIG. 5-6

| 41 | 38 | 31 | 22 | 21 | 15 | 11 | 9 | 4 | 3 |

FIG. 5-7

The techniques for ordering values in various desired sequences will be discussed in Chapter 12, Sorting.

It is often desirable to group two or more lists to accomplish some given purpose. Suppose for example, we have three related lists,

C

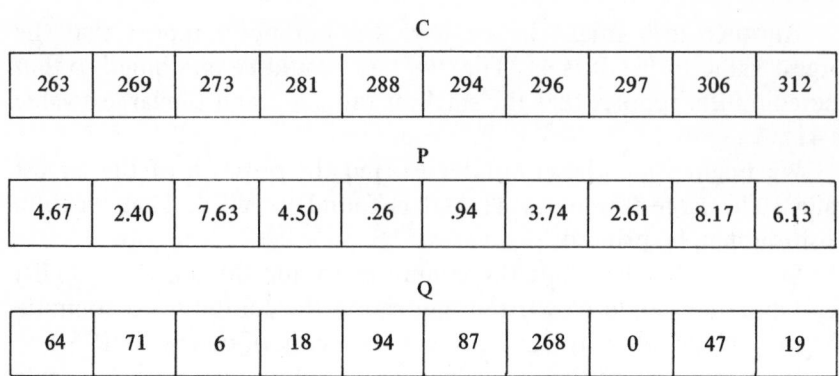

| 263 | 269 | 273 | 281 | 288 | 294 | 296 | 297 | 306 | 312 |

P

| 4.67 | 2.40 | 7.63 | 4.50 | .26 | .94 | 3.74 | 2.61 | 8.17 | 6.13 |

Q

| 64 | 71 | 6 | 18 | 94 | 87 | 268 | 0 | 47 | 19 |

FIG. 5-8

C, P, and Q (Fig. 5-8). Assume that list C contains ten catalog numbers of ten parts used in Super-8 Autos. List P contains corresponding selling prices for the ten parts and list Q contains the quantities-on-hand of those parts.

We may assume that the three lists are synchronized. That is, C_1, P_1, and Q_1 are related. This means that if a person wishes to know what the selling price and the quantity-on-hand are of part 263, all he has to do is note the position of 263 in list C and then jot down the contents of the first position of lists P and Q. Similarly, if he wishes to know what the selling price and the quantity-on-hand are of part 296, he first observes that catalog number 296 is in the seventh cell of list C. Then he writes down the seventh values found in lists P and Q.

It is simple for a person to look down a list of ten values to find the particular value he is looking for and then find corresponding values in two or more additional lists. Where thousands of items are involved, though, and answers must be quickly obtained, it is not practical to perform the work manually. A computer may be used to provide the immediate responses required.

A file containing many thousands of Super-8 Auto parts, along with their prices and quantities-on-hand, may be located at a central warehouse. The stockkeeper at the warehouse may be able to interrogate the file by means of a teletype located at the side of his desk. When he receives a phone call from a distant service shop requesting information about a part, he turns to the teletype, keys in an inquiry, and receives an answer within seconds. If the part desired is available, the stockkeeper is able to tell how many of the part are in stock and is able to quote a price.

When he hangs up the phone, another request may come in from another service station requesting information about one or more other parts. With the aid of the teletype, the stockkeeper is able to answer this and further inquiries, thus helping to satisfy Super-8 Auto customers.

The file at the central warehouse must, of course, be kept up to date; otherwise erroneous information might be given by the stockkeeper. It is the responsibility of the personnel at the warehouse to record regularly the parts received and those disbursed. The updating of the file must be done frequently enough so that most Super-8 Auto customers are given correct information most of the time. This may require that the central file be updated weekly, daily, or even continuously depending on the quality of the information the warehouse wants to give. Updating infrequently is less expensive but may result in creating too many unhappy Super-8 Auto customers. Updating continuously may be highly desirable but too expensive to be feasible. Some compromise must be found by those who administer the warehouse.

In Chapter 11 we shall discuss some of the methods that a computer uses when it attempts to find a given value in a list.

Multidimensional arrays are used in much the same way that lists are used. In fact, a multidimensional array is nothing more than a series of lists grouped as a single unit. You will recognize the values shown in the 3 X 10 array in Fig. 5-9. The name of the array is S. It has three rows and ten columns. Referring to an earlier example

S

263	269	273	281	288	294	296	297	306	312
4.67	2.40	7.63	4.50	.26	.94	3.74	2.61	8.17	6.13
64	71	6	18	94	87	268	0	47	19

FIG. 5-9

where three lists were used, C, P, and Q, you can see that the first row contains ten catalog numbers, the second row contains ten corresponding prices, and the third row contains ten corresponding quantities-on-hand.

When referencing a particular cell of a two-dimensional array, two subscripts are needed. Thus, $S_{1,3}$ refers to the contents of a single cell in the S array—the one located in the *first row, third column*. A glance at the array shows that the contents of that cell concerns Super-8-Auto part catalog number 273.

A two-part subscript always gives the row first and then the column. Thus the array value $S_{2,3}$ gives the price corresponding to Super-8 Auto part 273 (7.63) and $S_{3,3}$ gives the quantity-on-hand of that part (6).

Arrays may, of course, have more than two dimensions. An example of a three-dimensional array, W, is given in Fig. 5-10. The W array is a 3 X 4 X 5 array (three rows, four columns, and five pages). It represents the work records of four men on five days Monday through Friday. On Monday, man 1 produced 16 gears, 4 shafts, and 6 axles; man 2 produced 13 gears, 8 shafts, and 4 axles; etc. (Columns on each page represent man numbers; rows 1, 2, and 3 represent gears, shafts, and axles, respectively.)

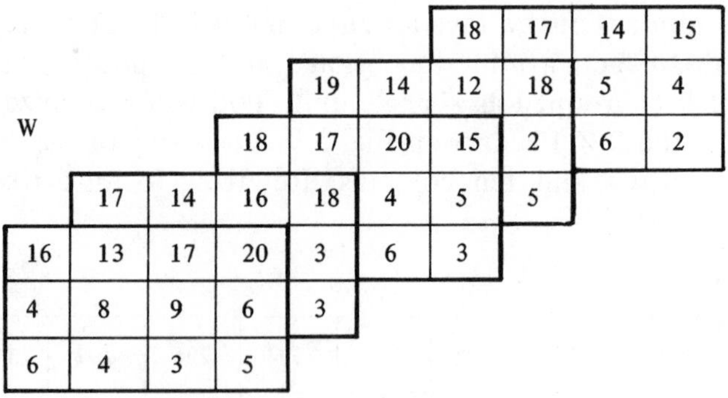

```
                              18   17   14   15
                         19   14   12   18    5    4
   W                18   17   20   15    2    6    2
            17  14  16   18    4    5    5
        16  13  17   20    3    6    3
        4    8   9    6    3
        6    4   3    5
```

FIG. 5-10

On Tuesday, man 1 produced 17 gears. We cannot see how many shafts and axles he produced because the figures are hidden behind Monday's report.

You can see that the W array needs three-part subscripts to identify any given cell. Thus $W_{2,4,3}$ represents the information given in the second row and fourth column of the third working day. We interpret this to mean that man 4 produced five shafts on Wednesday.

If we wish to, we may expand the three-dimensional array given above to a four-dimensional one, where the subscripts have four parts. Thus $W_{3,2,4,4}$ might represent the cell that tells how many axles man 2 built on Thursday during the fourth week of the month. There is no limit to how many dimensions an array may have providing the designer of the array is able to assign a meaning to each dimension.

EXERCISES

5-1. What is the name given to a single-dimensional array? To a two-dimensional array?

5-2. What is meant by the *size* of the list?

5-3. Define *subscript.*

5-4. What is meant by the term *sorting*?

5-5. How many dimensions may arrays have?

Chapter 6

SET THEORY

A *set* S is a collection of objects. The objects are called elements or members of S. For example, S may consist only of the three numbers 1, 2, and 3. Therefore, S may be defined by listing its elements within braces by writing '

$$S = \{1,2,3\}$$

Similarly, set T consisting of the elements a, b, c, d, and e is defined by writing

$$T = \{a,b,c,d,e\}$$

If we say that x is a member of T by writing

$$x \in T$$

we mean that x is a, b, c, d, or e. If we say that x is not a member of T by writing

$$x \notin T$$

we mean that x is not a, b, c, d, or e.

Sometimes the method of defining a set by exhibiting its elements within braces is unsatisfactory. This is when a set consists of a great many elements. For example, consider the set B consisting of all the positive integers from 101 to 200. Clearly, it would be cumbersome to list 100 integers within braces to define B. Therefore it is necessary to develop another method to define a set.

Let P be the set of positive integers. Therefore examples of valid relationships concerning set P are

$$5 \in P$$
$$-10 \notin P$$
$$1 \in P$$
$$1/3 \notin P$$
$$200 \in P$$
$$.6 \notin P$$

Hence by writing

$$B = \{x; x \in P, 100 < x < 201\}$$

we define the members of B to be integers ranging between 101 and 200.

Consider set D, defined by writing

$$D = \{a; a \in P, a < 7\}$$

which is clearly equivalent to

$$D = \{1,2,3,4,5,6\}$$

Furthermore, we may now define set S another way by writing

$$S = \{y; y \in P, y < 4\}$$

EXERCISE

6-1. Provide an alternative way of defining the following sets:

(a) $A = \{4,5,6,7\}$

(b) $B = \{20,21,22,23,24\}$

(c) $C = \{x; x \in P, 10 < x < 20\}$

(d) $D = \{y; y \in P, 1000 < y < 1002\}$

Elements of sets are not limited to the realm of numbers and abstract characters. Sets may be comprised of people. For example, set H may consist of the students at Harvard University. H may be defined by writing

$$H = \{x; x \text{ is a student at Harvard University}\}$$

Or, set C may consist of the Chicago Bears football team. C may be defined by writing

$$C = \{y; y \text{ is a Chicago Bear football player}\}$$

If a set N consists of no members, it is empty and defined as a *null* set. A null set N is defined by writing

$$N = \{\emptyset\}$$

N is a member of every set S.

A *universal* set U consists of all the possible elements that might be considered in a particular situation. The set P of positive integers is an example of a set that is often used as a universal set. P may be defined by writing

$$P = \{1,2,3,4, \ldots\}$$

with the three dots indicating that P consists of all the positive integers.

Consider the set Q where Q is the set of all even positive integers. This is shown by writing

$$Q = \{2,4,6,8, \ldots\}$$

with the three dots indicating that all even positive integers are included in this set.

Another method of defining set Q might be by writing

$$Q = \{x; x \in P, x \text{ divisible by 2}\}$$

which means that Q consists of all elements x having the properties that x is a positive integer and is divisible by 2. Clearly, Q is the set of all even positive integers.

Every member of Q is a member member of P. Therefore Q is a *subset* of P, written

$$Q \subset P$$

Returning to set $S = \{1,2,3,\}$, we see that every element of S is in P; therefore

$$S \subset P$$

Also, since 1 and 3 are not members of Q written

$$1 \notin Q, \quad \text{and} \quad 3 \notin Q$$

we see that S is not a subset of Q, written

$$S \not\subset Q$$

Finally, let set T be defined as

$$T = \{2,4,6,8\}$$

Now the following relationships are valid:

$$T \subset Q \subset P$$

which implies that

$$T \subset P$$

The records of a master file constitute a set. The fields that make up the master record form a set. Set M, a personnel master file, may be defined by writing

$$M = \{x; x \text{ is a personnel master record}\}$$

Or, a personnel master record may be defined as set R by writing

$$R = \{y; y \text{ is a personnel record field}\}$$

If the number of data fields is small, it may be convenient to write

$$R = \{\text{NAME, ADDRESS, JOB CLASS, SALARY, BIRTHDAY, SEX}\}$$

And, of course,

$$T = \{\text{NAME, ADDRESS}\}$$

would be a subset of R, so that we could write

$$T \subset R$$

The *complement* of a set S, written S', consists of the elements in the universal set U for S minus the elements in S. Therefore, letting U be P, the set of positive integers, we say that

$$U' = \{\emptyset\}$$
$$\{\emptyset\}' = U$$

Furthermore, by defining S to be

$$S = \{1,2,3\}$$

we see that

$$S' = \{x; x \in P, x > 3\}$$

or, in another way,

$$S' = \{4,5,6,7, \ldots\}$$

To reinforce the idea of a complement of a set let us define set $P = \{1,2,3,4, \ldots\}$, where P is U, set $Q = \{2,4,6,8, \ldots\}$, set $R = \{1,3,5,7, \ldots\}$, set $M = \{x; x \in P, x < 6\}$, and set $N = \{x; x \in P, 0 < x < 101\}$. Therefore

$$R' = Q$$

$$Q' = R$$

$$U' = P' = \{\emptyset\}$$

$$M' = \{6,7,8,9, \ldots\} = \{x; x \in P, x > 5\}$$

$$N' = \{101,102,103,104, \ldots\} = \{x; x \in P, x > 100\}$$

and

$$R \subset U$$

$$Q \subset U$$

$$M \subset N \subset U$$

$$R' \subset U$$

$$Q' \subset U$$

$$N' \subset M' \subset U$$

The *union* of two or more sets, S_1 to S_n, is another set S_t, which consists of all the elements in sets S_1 to S_n. (The subscripts are nothing more than a convention used to distinguish between the sets.) For example, let

$$A = \{1,2,3,4,5\}$$

$$B = \{3,4,5,6,7\}$$

Therefore A union B, written $A \cup B$, is the set

$$A \cup B = \{1,2,3,4,5,6,7\} = \{x; x \in P, 0 < x < 8\}$$

Furthermore, letting the universal set U be the set of positive integers and sets Q and R be the sets of even and odd integers, respectively, we also see that

$$Q \cup R = U$$

$$A \cup U = U$$

$$B \cup U = U$$

The *intersection* of two or more sets is a set consisting of those elements that are common to all sets. For example, let

$$A = \{1,2,3,4,5\}$$

$$B = \{3,4,5,6,7\}$$

Therefore A intersection B, written $A \cap B$, is the set

$$A \cap B = \{3,4,5\}$$

If two or more sets have no common elements, they are *disjoint* and their intersection is the null set. For example, let

$$C = \{1,2,3\}$$

$$D = \{4,6,8\}$$

Hence

$$C \cap D = \{\emptyset\}$$

Also,

$$A \cap B \cap C = \{3\}$$

$$A \cap C = \{1,2,3\} = C$$

$$A \cap D = \{4\}$$

$$B \cap C = \{3\}$$

$$B \cap D = \{4,6\}$$

$$A \cap B \cap C \cap D = \{\emptyset\}$$

EXERCISES

6-2. Let set $P = \{1,2,3,4,\ldots\}$, set $A = \{1,2,6,8\}$, set $B = \{6,8\}$, set $C = \{x; x \in P, 0 < x < 9\}$, and set $D = \{x; x \in P, 6 < x < 8\}$. Discuss the relationships among the sets.

6-3. If set $A = \{x,y,z\}$, set $B = \{a,b,x\}$, and set $C = \{b,z\}$, determine each of the following:

(a) $A \cap B$

(b) $A \cap C$

(c) $B \cap C$

(d) $A \cup B$

(e) $A \cup C$

(f) $B \cup C$

6-4. Let P be the set of positive integers and let

$$A = \{x; x \in P, x < 20\}$$
$$B = \{y; y \in P, y > 10\}$$
$$C = \{z; z \in P, z \text{ divisible by } 3\}$$

Determine each of the following:

(a) $A \cap B$

(b) $A \cap C$

(c) $B \cap C$

(d) $A \cup B$

(e) $A \cup C$

(f) $B \cup C$

(g) $A \cup B \cup C$

(h) $A \cap B \cap C$

6-5. Let U be the set of positive integers and find the complements of the following:

(a) $A = \{5,6,7,8\}$

(b) $B = \{2,3,4,5,\ldots\}$

(c) $C = \{\emptyset\}$

(d) $D = \{x; x \in P, x > 5\}$

(e) $E = \{5,10,15,20, \ldots\}$

6-6. Exhibit the different subsets of a set with two elements. Of a set with three elements. Of a set with four elements. What relationship do you observe between the number of elements in a set and the number of subsets of a set?

6-7. Let sets A and B be defined as in 6-2, set $C = \{x; x \in P$ and $x > 5\}$, and set $D = \{x; x \in P$ and, $x < 8\}$.
Define the sets resulting from the following unions:

(a) $A \cup C$

(b) $A \cup D$

(c) $B \cup C$

(d) $B \cup D$

(e) $C \cup D$

(f) $A \cup B \cup C$

(g) $A \cup B \cup C \cup D$

6-8. Let sets A, B, C, and D be defined as above. Set $E = \{1,3,5,7\}$ and set $F = \{4\}$. Define the sets resulting from the following intersections:

(a) $A \cap E$

(b) $E \cap F$

(c) $A \cap C \cap F$

(d) $A \cap B \cap F$

(e) $D \cap F$

(f) $A \cap B \cap C \cap D \cap E \cap F$

A very useful tool to demonstrate relationships between sets is the *Venn diagram*. For example, let the universal set U be the set of positive integers, set $R = \{1,2,3,4\}$, set $S = \{3,4,5,6\}$, set $T = \{5,6,7,8\}$, and set $W = \{2,3\}$. The following illustrations demonstrate the function of a Venn diagram.

The shaded portions of the diagrams represent the sets $R \cup S$, $R \cap S$, and $R \cup S \cup T$, respectively.

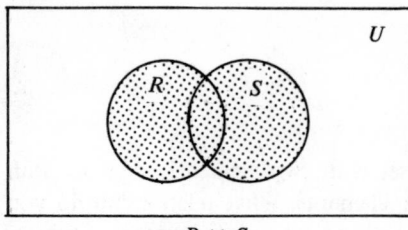

$R \cup S$

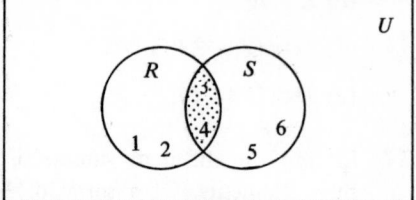

$R \cap S$

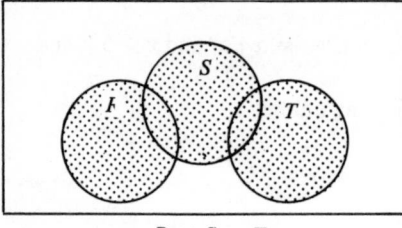

$R \cup S \cup T$

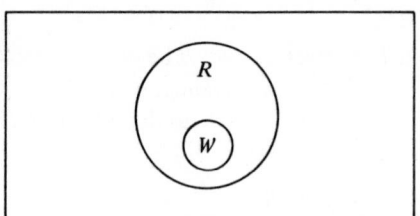

$W \subset R$

EXERCISES

6-9. Let set P be the universal set in all cases, and set $A = \{x; x \in P, x < 7\}$, set $B = \{y; y \in P, y < 10\}$, and set $C = \{z; z \in P, z$ divisible by $3\}$. Prepare Venn diagrams to demonstrate the following sets:

(a) $A \cap B$

(b) $A \cap C$

(c) $B \cap C$

(d) $A \cup B$

(e) $A \cup C$

(f) $B \cup C$

(g) $A \subset B$

6-10. Let set P be the universal set, set $A = \{1,2,3,4\}$, set $B = \{5,6\}$, set $C = \{3,4,5\}$, and set $D = \{1,2\}$.

Prepare Venn diagrams to demonstrate the following sets:

(a) $D \subset A$

(b) $A \cup B$

(c) $A \cap B$

(d) $A \cup C$

(e) $A \cap C$

(f) $A \cup D$

(g) $A \cap D$

(h) $B \cup C$

(i) $B \cap C$

(j) $B \cup D$

(k) $B \cap D$

(l) $C \cup D$

(m) $C \cap D$

(n) $A \cup B \cup C$

(o) $A \cup B \cup D$

(p) $B \cup C \cup D$

(q) $A \cup C \cup D$

(r) $A \cap B \cap C$

(s) $A \cap B \cap D$

(t) $B \cap C \cap D$

(u) $A \cap C \cap D$

(v) $A \cup B \cup C \cup D$

(w) $A \cap B \cap C \cap D$

Let set $A = \{a,b,c\}$, set $B = \{b,c,d\}$, and set $C = \{e\}$. Consider the set $(A \cap B) \cup C$. Is this the same as set $A \cap (B \cup C)$? Consider $(A \cap B) \cup C$. Performing the operations within the parentheses first, we have

$$A \cap B = \{b,c\}$$

and

$$(A \cap B) \cup C = \{b,c,e\}$$

Now considering $A \cap (B \cup C)$ and also performing the operations within the parentheses first, we have

$$B \cup C = \{b,c,d,e\}$$

and

$$A \cap (B \cup C) = \{b,c\}$$

Therefore we find that

$$(A \cap B) \cup C \neq A \cap (B \cup C)$$

and we can say that the order of performing set operations is critical with the performance of operations within parentheses first essential. Furthermore, if there are nested parentheses, the operation within the innermost parentheses is processed first. Therefore

$$((A \cup B) \cup C)$$

means that $A \cup B$ will be developed first and then the union of set $(A \cup B)$ and set C will be performed.

Assume that it is necessary to prepare a statistical report for those female students who are in the schools of education or business. Define the sets by writing

$$S = \{x; x \text{ is a student}\}$$

$$A = \{x; x \in S, x \text{ is a female}\}$$

$$B = \{x; x \in S, x \text{ is in the school of education}\}$$

$$C = \{x; x \in S, x \text{ is in the school of business}\}$$

We are interested in the set defined by

$$T = (A \cap B) \cup (A \cap C)$$

T is shown by the shaded area in the following Venn diagram.

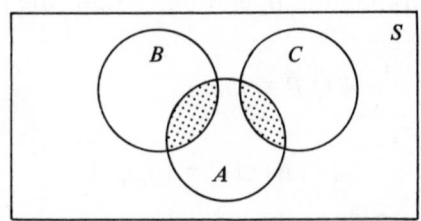

EXERCISE

6-11. Let sets A, B, C, and D be defined as in Exercise 6-10. Perform the following set operations:

    (a) $(A \cup B) \cap (C \cup D)$

    (b) $(A \cap B) \cup (C \cap D)$

    (c) $(A \cup B \cup C) \cap D$

    (d) $(A \cap B \cap C) \cup D$

    (e) $(A \cap B) \cup (A \cap C) \cup (A \cap D)$

    (f) $(A \cup B) \cap (A \cup C) \cap (A \cup D)$

Example 1. As another example, assume that we have given an employee master record with the following data elements:

    1. Employee name.

    2. Employee number.

    3. Job class.

    4. Birthday.

    5. Sex.

    6. Marital status.

    7. Rate of pay.

We are interested in determining those employees who are single male engineers over 50 years of age who make in excess of $15,000 per year. The pertinent sets may be defined by writing

    $U = \{x; x \text{ is an employee}\}$

    $A = \{x; x \in U, x \text{ is single}\}$

    $B = \{x; x \in U, x \text{ is male}\}$

    $C = \{x; x \in U, x \text{ is an engineer}\}$

    $D = \{x; x \in U, x \text{ is over 50 years old}\}$

    $E = \{x; x \in U, x \text{ has a salary in excess of \$15,000 per year}\}$

Clearly, we are interested in those employees who are in the set defined by the intersection of sets *A, B, C, D,* and *E.* Or,

$$A \cap B \cap C \cap D \cap E$$

and this intersection is shown as the shaded area in the following Venn diagram.

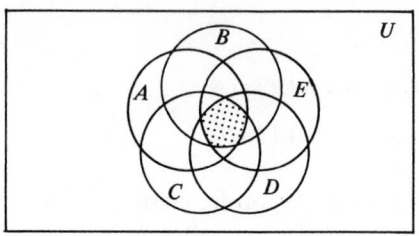

EXERCISE

6-12. How would you proceed to find the set of those single male employees who are either managers or engineers in excess of 40 years of age? Prepare a Venn diagram to show this set.

Chapter 7

DECISION TABLES

A decision table shows what action is to be taken for various conditions that may occur in an application. IF certain conditions are true, THEN certain action is necessary. An illustration might serve to convey the concept of a decision table. Consider an inventory control application in which it is necessary to maintain stock balances. We must consider appropriate action for issue and receipt transactions. The transactions represent the conditions that may occur. Figure 7-1 represents a decision table that defines the processing of the inventory control transactions. The only conditions to be considered are whether or not the transaction is an issue or receipt. If the transaction is an issue, the necessary action is a reduction of the stock balances. If the transaction is a receipt, the action to be taken is an increasing of the stock balances. Appropriate COBOL sentences to process the information in the table are shown in the partial program of Fig. 7-2.

TRANSACTION IS ISSUE	Y	N
TRANSACTION IS RECEIPT	N	Y
ADD QUANT TO OH-BAL	–	X
SUB QUANT FROM OH-BAL	X	–

ELEMENTS OF A DECISION TABLE

FIG. 7-1

IBM

COBOL Coding Form

SYSTEM				PUNCHING INSTRUCTIONS			PAGE	OF
PROGRAM			GRAPHIC			CARD	*	
PROGRAMMER	DATE		PUNCH			FORM #	IDENTIFICATION 73 80

```
SEQUENCE   A  B                        COBOL STATEMENT
(PAGE)(SERIAL) CONT
01
02
03   FILE  SECTION.
04
05   Ø1  INVENTORY-TRANSACTION.
06       Ø2  TRANSACTION-TYPE      PICTURE  XX.
07           88  ISSUE            VALUE  '1'.
08           88  RECEIPT          VALUE  '66'.
09       Ø2  PRODUCT-NUMBER-T     PICTURE  X(1Ø).
10       Ø2  QUANTITY             PICTURE  S9(5).
11
12
13
14   Ø1  INVENTORY-MASTER.
15       Ø2  PRODUCT-NUMBER-M     PICTURE  X(1Ø).
16       Ø2  DESCRIPTION          PICTURE  X(2Ø).
17       Ø2  ON-HAND-BALANCE      PICTURE  9(5).
18       Ø2  RE-ORDER-POINT       PICTURE  9(5).
19       Ø2  RE-ORDER-QUANTITY    PICTURE  9(5).
20
21
22
23
24
```

*A standard card form — IBM Electro (6189) 7, is available for punching source statements from this form.
Instructions for using this form are given in any IBM COBOL reference manual
Address comments concerning this form to IBM Corporation, Programming Publications, 1271 Avenue of the Americas, New York, New York 10020

FIG. 7-2

IBM COBOL Coding Form

SYSTEM | PUNCHING INSTRUCTIONS | PAGE OF
PROGRAM | GRAPHIC | PUNCH | CARD FORM # | * | IDENTIFICATION 73 80
PROGRAMMER | DATE

```
SEQUENCE   A   B
01
02
03      PROCEDURE DIVISION.
04
05
06      UPDATE.
07          IF ISSUE COMPUTE ON-HAND-BALANCE = ON-HAND-BALANCE - QUANTITY
08          IF RECEIPT COMPUTE ON-HAND-BALANCE = ON-HAND-BALANCE +
09              QUANTITY.
10
11
12
13
14
15
16
17
18
19
20
21
22
23
24
```

*A standard card form, IBM Electro C61897, is available for punching source statements from this form.
Instructions for using this form are given in any IBM COBOL reference manual.
Address comments concerning this form to IBM Corporation, Programming Publications, 1271 Avenue of the Americas, New York, New York 10020.

FIG. 7-2 (cont.)

CONDITION STUB	CONDITION ENTRY
ACTION STUB	ACTION ENTRY

FIG. 7-3   Elements of a decision table.

A decision table is a tool for the data processor. It enables a systems analyst to readily convey important decision-making information to a programmer or to a reader of a set of systems specifications. What is of utmost importance is the fact that an individual without a background in data processing can read a decision table with a great deal of comprehension. By the same token, a programmer can prepare a program without the necessity of preparing a program flowchart.

A decision table is divided into four parts, as shown in Fig. 7-3. The parts consist of a condition stub and entries and an action stub and entries. The condition stub sets forth the conditions that may exist and the action stub outlines particular actions to be taken for combinations of condition entries.

It would not be difficult for one to reflect a moment and notice a correspondence between a decision table and a program flowchart. For instance, consider the problem of determining what to wear to school. This is obviously not a typical data processing application. But this is perfectly acceptable since we are merely attempting to become acclimated to decision table concepts that involves clear, concise thinking. This problem is solved by both the decision table of Fig. 7-4,† and the program flowchart of Fig. 7-5.‡ For example, reading the first column of the decision table, we see that if it is raining and the temperature is over 70 degrees the following action is necessary:

†Awad, Elias M., *Business Data Processing*, 2nd Ed. Englewood Cliffs, N. J.: Prentice-Hall, Inc. (1968), p. 411.

‡*Ibid.*, p. 409.

CONDITION ENTRY ENCOUNTERED

	Alternative one	Alternative two	Alternative three	Alternative four
Raining and temp over 70°	Y	N	N	N
Forecast clear and temp under 70°	--	Y	N	N
Forecast rain and temp under 70°	--	--	Y	N
Wear light clothes	X	--	--	--
Wear regular clothes	--	X	X	--
Wear summer clothes	--	--	--	X
Wear a sweater	--	X	X	--
Wear a raincoat	X	--	--	--
Carry a raincoat	--	--	X	--
Carry an umbrella	X	--	X	--
Wear overshoes	X	--	X	--
Go to car	X	X	X	X

IF, or (condition) — first three rows
THEN, or (action) — remaining rows

ACTION ENTRY

Abbreviations

Y = yes
N = no
-- = irrelevant action (or condition)
X = completion of action statement

FIG. 7-4 Decision table to determine what to wear to school.

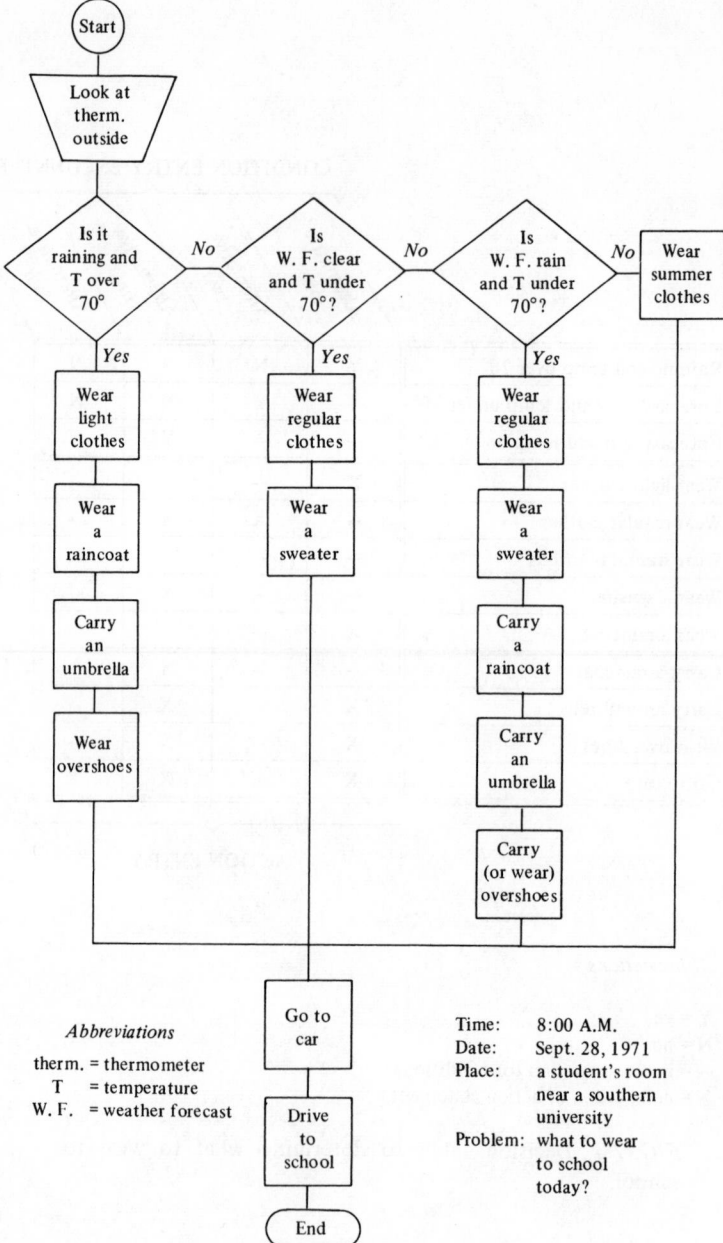

FIG. 7-5 Systems and procedures flowchart to determine what to wear to school.

1. Wear light clothes.
2. Wear a raincoat.
3. Carry an umbrella.
4. Wear overshoes.
5. Go to car.
6. Drive to school.

Looking at the second column, we see that if the forecast is clear and temperature is under 70 degrees the only actions consist of

1. Wear regular clothes.
2. Wear a sweater.
3. Go to car.
4. Drive to school.

Finally, notice that different conditions may require one or more of the same actions. For example, all the conditions require the actions

1. Go to car.
2. Drive to school.

Both of the above data processing tools require basically the same logical thinking. Furthermore, they demonstrate that this same logical thinking may also be used in many facets of our everyday life. For instance, we could prepare a decision table for a two-week vacation to Cape Cod, or we could prepare a decision table for a woman's journey into a supermarket, working with a limited budget.

Rule No.	1	2	3	4
Credit Limit OK	Y	N	N	N
Pay Experience Good		Y	N	N
Special Clearance Obtained			Y	N
Approve Order	X	X	X	
Return to Sales				X

FIG. 7-6 Limited-entry table.

Example 1. Figure 7-6† is another example of a decision table. This

†Fisher, F. Peter, and Swindle, George F., *Computer Programming Systems.* New York: Holt, Rinehart and Winston, Inc. (1964), p. 49.

IBM COBOL Coding Form

SYSTEM		PUNCHING INSTRUCTIONS			PAGE OF
PROGRAM		GRAPHIC		CARD FORM # *	IDENTIFICATION
PROGRAMMER	DATE	PUNCH			73 80

```
SEQUENCE   A   B                                  COBOL STATEMENT
01
02        FILE SECTION.
03
04        Ø1  CUSTOMER-MASTER.
05            Ø2  CUSTOMER-NO-A           PICTURE X(24).
06
07
08
09
10            Ø2  CREDIT-CODE             PICTURE X.
11                88  CREDIT-LIMIT-OK         VALUE '1'.
12                88  CREDIT-LIMIT-UNSAT      VALUE '2'.
13            Ø2  PAY-EXPERIENCE-CODE     PICTURE X.
14                88  PAY-EXPERIENCE-GOOD         VALUE '1'.
15                88  PAY-EXPERIENCE-BAD          VALUE '2'.
16
17
18        Ø1  ORDER-TRANSACTION.
19            Ø2  CUSTOMER-ON-B           PICTURE X(14).
20
21            Ø2  CLEARANCE-CODE          PICTURE X.
22                88  SPECIAL-CLEARANCE-OBTAINED      VALUE '1'.
23                88  SPECIAL-CLEARANCE-NOT-OBTAINED  VALUE '2'.
24
```

*A standard card form, IBM Electro C61897, is available for punching source statements from this form.
Instructions for using this form are given in any IBM COBOL reference manual.
Address comments concerning this form to IBM Corporation, Programming Publications, 1271 Avenue of the Americas, New York, New York 10020.

FIG. 7-7

IBM COBOL Coding Form

SYSTEM | PUNCHING INSTRUCTIONS | PAGE OF
PROGRAM | GRAPHIC | CARD FORM # | * IDENTIFICATION 73 80
PROGRAMMER | DATE | PUNCH

SEQUENCE (PAGE) (SERIAL) | CONT. | A | B | COBOL STATEMENT

```
01
02
03  PROCEDURE DIVISION.
04
05
06  ORDER-ACTION-ROUTINE.
07      IF CREDIT-LIMIT-OK
08          OR
09      IF CREDIT-LIMIT-UNSAT AND IF PAY-EXPERIENCE-GOOD
10          OR
11      IF CREDIT-LIMIT-UNSAT AND IF PAY-EXPERIENCE-BAD AND IF
12          SPECIAL-CLEARANCE-OBTAINED GO TO ACCEPT-ORDER-ROUTINE
13          ELSE
14      GO TO REJECT-ORDER-ROUTINE.
15
16
17
18
19
20
21
22
23
24
```

*A standard card form, IBM Electro C61897, is available for punching source statements from this form.
Instructions for using this form are given in any IBM COBOL reference manual.
Address comments concerning this form to IBM Corporation, Programming Publications, 1271 Avenue of the Americas, New York, New York 10020

FIG. 7-7 (cont.)

decision table depicts the steps required in determining whether or not a customer is to be permitted credit. The condition stub includes the following items: (1) credit limit OK, (2) pay experience good, and (3) special clearance obtained. The action stub includes either an approve or disapprove response. If the credit limit is satisfactory, the order is to be approved. If the credit limit is unsatisfactory but the pay experience is good, the order is to be approved. If the credit limit is not OK, the pay experience is bad, but special clearance is obtained, then the order is to be approved. And, finally, if the credit limit is unsatisfactory, the pay experience is not good, and no special clearance is obtained, the order is to be disapproved. Figure 7-7 represents COBOL routines to process this decision table. Two records are used in this program. First, a customer master record, which contains previous pay experience and the credit limit, and second, the order transaction record, which contains the clearance code.

Example 2. Figure 7-8† is a decision table for a life insurance application. The problem is to determine insurance rates and policy limits based upon various parameters. There will be 30 rules in the table. The rules will be functions of age, health, and the section of the country for a prospective insuree. Based upon combinations of the conditions, appropriate actions will consist of an insurance rate per 1000 and a policy limit. For example, if the individual is between 25 and 34 years of age, in excellent health, and is from the east, the action consists of issuing a policy at a rate per $1000 of $1.57 up to

	Rule 1	Rule 2			Rule 30
Age	25-34	25-34			65 and over
Health	Excellent	Excellent			Poor
Section of Country	East	West			West
Rate per 1000	1.57	1.72			5.92
Policy Limit	200,000	200,000			20,000

FIG. 7-8   Life insurance decision process.

†*Ibid.,* p. 488.

a limit of $200,000. Or, if the individual is 65 years of age or older, in poor health, and is from the west, his rate is $5.92 per $1000 of coverage up to a limit of $20,000.

Example 3. Let us now expand our inventory control application. We said that issue and receipt transactions were used in maintaining stock balances. Additionally, however, if the on-hand balance falls below a safety level, defined by the reorder point (ROP), order action is necessary. This table is reflected in Fig. 7-9. Issues from stock and receipts into stock are processed resulting in a new stock balance. For issue transactions, a query must also be made to determine those stock balances that are equal to or fall below the ROP. At that time some form of order action must be initiated to ensure that customer orders are satisfied.

TRANSACTION IS ISSUE	YES	YES	NO	NO
TRANSACTION IS RECEIPT	NO	NO	YES	YES
OH BALANCE > ROP	YES	NO	–	–
ADD QUANT TO OH BAL	–	–	X	X
SUB QUANT FROM OH BAL	X	X	–	–
ORDER ACTION REQ'D	–	X	–	–

FIG. 7-9

Example 4. A large university has implemented a student information system. Each record of the student master file consists of fields pertaining to the student's personal data, financial data, and grade history. Assume that it is necessary to prepare a statistical report, as shown in Fig. 7-10, for the government. This report lists the numbers of full-time and part-time male and female students for both the day and evening schools. A decision table to logically process each of the desired catagories is shown in Fig. 7-11. For example, the first column defines what processing is necessary for a record of a full-time (FT) male day student. That is, if a record is such, the action necessary consists of:

MALE FULL TIME (FT)DAY	1250
MALE FULL TIME (FT) EVENING	300
TOTAL MALE FULL TIME (FT)	1550

FEMALE FULL TIME (FT) DAY	900
FEMALE FULL TIME (FT) EVENING	150
TOTAL FEMALE FULL TIME (FT)	1050

TOTAL FULL TIME (FT) STUDENTS DAY	2150
TOTAL FULL TIME (FT) STUDENTS EVENING	450
TOTAL FULL TIME (FT) STUDENTS	2600

MALE PART TIME (PT) DAY	600
MALE PART TIME (PT) EVENING	1450
TOTAL MALE PART TIME (PT)	2050

FEMALE PART TIME (PT) DAY	250
FEMALE PART TIME (PT) EVENING	500
TOTAL FEMALE PART TIME (FT)	750

TOTAL PART TIME (PT) STUDENTS DAY	850
TOTAL PART TIME (PT) STUDENTS EVENING	1950
TOTAL PART TIME (PT) STUDENTS	2800

TOTAL STUDENTS	5400
TOTAL MALE STUDENTS	3600
TOTAL FEMALE STUDENTS	1800

FIG. 7-10   Enrollment statistics.

1. Incrementing male FT day counter.
2. Incrementing total male FT counter.
3. Incrementing total FT student day counter.
4. Incrementing total FT student counter.
5. Incrementing total student counter.
6. Incrementing total male student counter.

IF									
MALE		Y	Y	Y	Y	N	N	N	N
FEMALE		N	N	N	N	Y	Y	Y	Y
FT		Y	Y	N	N	Y	Y	N	N
PT		N	N	Y	Y	N	N	Y	Y
DAY		Y	N	Y	N	Y	N	Y	N
EVENING		N	Y	N	Y	N	Y	N	Y
THEN									
INCREMENT	MALE FT DAY COUNTER	X							
	MALE FT EVEN		X						
	TOTAL MALE FT	X	X						
	FEMALE FT DAY					X			
	FEMALE FT EVEN						X		
	TOTAL FEMALE FT					X	X		
	TOTAL FT STUD DAY	X				X			
	TOTAL FT STUD EVEN		X				X		
	TOTAL FT STUD	X	X			X	X		
	MALE PT DAY			X					
	MALE PT EVEN				X				
	TOTAL MALE PT			X	X				
	FEMALE PT DAY							X	
	FEMALE PT EVEN								X
	TOTAL FEMALE PT							X	X
	TOTAL PT STUD DAY			X				X	
	TOTAL PT STUD EVEN				X				X
	TOTAL PT STUD			X	X			X	X
	TOTAL STUD	X	X	X	X	X	X	X	X
	TOTAL MALE STUD	X	X	X	X				
	TOTAL FEMALE STUD					X	X	X	X

FIG. 7-11   Enrollment statistics decision table.

Example 5. In Examples 7 and 8 of Chapter 1, payroll calculations were performed. Let us develop a decision table that reflects the necessary logic required in such an application. In this example assume that it is necessary only to compute FICA at 5.4 percent of gross pay (see Exercise 7-4 for the expanded FICA calculation).

Four rules are required in the table shown in Fig. 7-12. Each of the four rules require the following calculations:

1. Compute gross-pay = std-pay + ovt-pay.
2. Compute non-tax-pay = (no-depend × 650)/52.

IF	R1	R2	R3	R4
CURRENT HOURS ⩽ 40	X	X		
CURRENT HOURS > 40			X	X
TAXABLE INCOME > 0	X		X	
TAXABLE INCOME ⩽ 0		X		X
THEN				
COMPUTE STD-PAY = HRS × RATE	Y	Y	N	N
STD-PAY = 40 × RATE	N	N	Y	Y
OVT-PAY = (CURRENT HOURS – 40) × (RATE × 1.5)	N	N	Y	Y
GROSS-PAY = STD PAY + OVT PAY	Y	Y	Y	Y
NON-TAX-PAY = (NO-DEPEND × 650)/52	Y	Y	Y	Y
TAX-PAY = GROSS-PAY – NON-TAX-PAY	Y	Y	Y	Y
FIT = .14 × TAX-PAY	Y	N	Y	N
FIT = 0	N	Y	N	Y
FICA = .054 × GROSS-PAY	Y	Y	Y	Y
▼ NET-PAY = GROSS-PAY – FIT-FICA	Y	Y	Y	Y

FIG. 7-12 Weekly payroll decision table.

3. Compute tax-pay = gross-pay – non-tax-pay.
4. Compute FICA = .054 × gross-pay.
5. Compute net-pay = gross-pay – FIT – FICA.

In addition, rule 1, for example, computes

1. Std-pay = hrs × rate.
2. FIT = .14 × tax-pay.

Notice that in the condition stub there are two questions to answer that result in four entries or rules. Exercise 7-4 requires the decision table to be expanded to ensure that FICA calculations be terminated after YTD earnings reach $8200.00. This additional condition will double the number of rules appearing in the condition entry portion of the decision table.

EXERCISES

7-1. A master file of the students at a university consists of the following data elements: name, address, age, sex code, and class code. Prepare a decision table to list a summary report of the 20-year-old freshman students. In the process, determine the number of records processed and the number of records found in the desired category.

7-2. Given the following step function, prepare a decision table to compute y depending on values of x:

$$y = y \qquad 0 \leqslant x < 10$$
$$y = x^2 \qquad 10 \leqslant x < 20$$
$$y = x^3 \qquad 20 \leqslant x$$

7-3. Prepare a program flowchart that corresponds to the decision table of Example 1. Of Example 3.

7-4. Expand the decision table of Example 5 to include interrogrations that suppress the FICA deduction once the $8200.00 gross earnings limit is reached.

Chapter 8

OPTIMIZING
BLOCKING
FACTORS

Magnetic Tape

Magnetic tape, along with the punched card and magnetic disk, is one of the principal medias utilized by Electronic Data Processing systems for storing data. Data are recorded on tape as magnetized spots called bits. Figure 8-1 pictorially illustrates how data reside on a small section of magnetic tape. In this particular example, 50 numeric, alpha, and special characters are shown. The characters are stored in parallel channels along the length of the tape, similar in concept to that of the punched card, although using a different coding system. The data placed on the tape can be retained indefinitely or erased by writing over it.

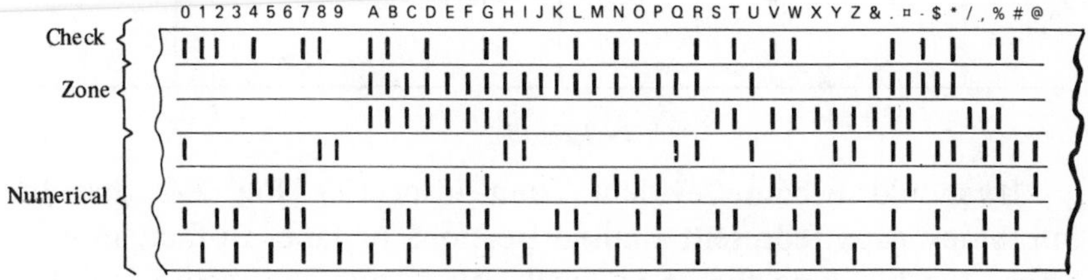

FIG. 8-1   Representation of magnetic tape showing stand-ard binary coded decimal interchange code.

The number of characters that can be placed on tape is a function of the tape density. Tape densities vary from 200 to 1600 characters per inch (CPI). Therefore a tape with a density of 200 CIP means that the 50 characters illustrated in Fig. 8-1 would fit in one quarter of an inch of tape. Even more startling, however, is the fact that the data stored in 20 punched cards may be placed on 1 inch of tape with a density of 1600 CPI. Since a reel of tape may be as much as 2400 feet in length it is possible to permit large numbers of records to be recorded on this media.

Figure 8-2 shows how records may be represented on a segment of tape. In the first example single records are separated by blank areas called interblock gaps. In the second example, the records are blocked four records per block, with the blocks separated by interblock gaps. The interblock gap (IBG), which varies in different EDP systems from .6 to .75 inches in length, is automatically produced at the end of each block or records during writing. A block of records may consist of one record or an integral number of records. In addition to separating blocks of records on the magnetic tape, the gap also allows time for starting and stopping the tape between processing of the record blocks.

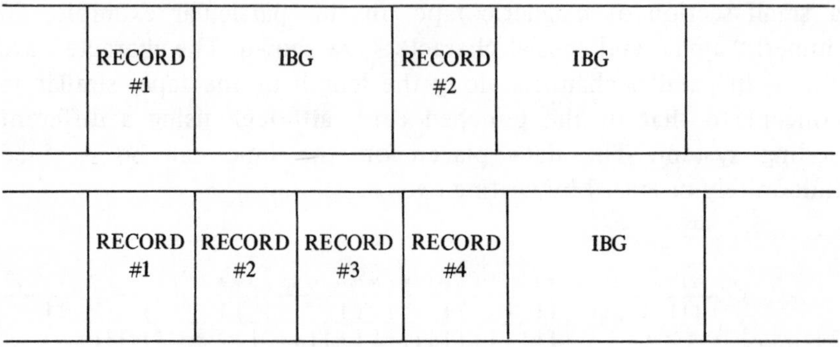

RECORD #1	IBG	RECORD #2	IBG	

RECORD #1	RECORD #2	RECORD #3	RECORD #4	IBG	

FIG. 8-2

It should become evident from inspecting Fig. 8-2 that the interblock gaps represent unused portions of tape. In fact, in cases where the blocking factor is small, say 1, and the record length is small, say 80 characters, most of the tape is devoted to gap areas. Therefore, one of the functions of the data processor is to develop a blocking factor that will maximize the tape storage capacity by minimizing the interblock gaps.

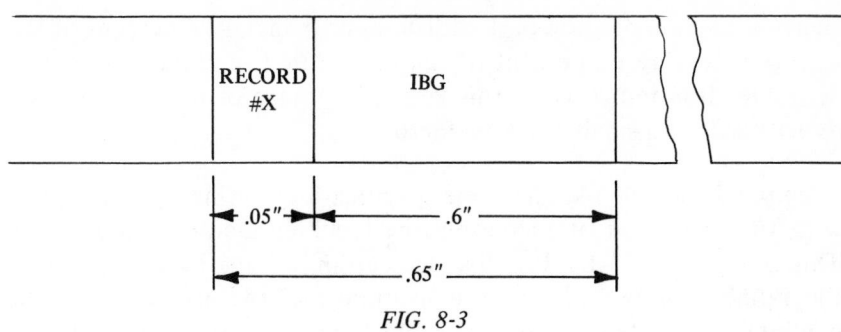

FIG. 8-3

Example 1. Let us determine the number of records that may be placed on a tape reel when the blocking factor is 1, as shown in Fig. 8-3. The specifications for this example consist of

1. Tape reel is 2500 feet or 30,000 inches in length.
2. Tape density is 1600 CPI.
3. IBG is .6 inches in length.
4. Data records are 80 characters long.
5. Blocking factor is 1.

Figure 8-3 demonstrates that the specifications of this example generate one record for each .65 inches of tape. Therefore on one reel of tape we have

30,000 inches/tape/.65 inches/record = 46,153 records/tape

There is no doubt that the placing of 46,153 records on a reel of tape is impressive and significantly reduces the amount of card handling required in a system. However, it should be readily noticeable that most of the tape is devoted to gap areas, since for every record that occupies .05 inches of tape there exists an IBG that requires .6 inches of tape.

Let us consider the tape usage efficiency in this example. We have

46,153 records/tape × .05 inches/record = 2,307.65 inches of
records/tape

and

46,153 records/tape × .6 inches/IBG = 27,691.80 inches of
IBG/tape

Hence a blocking factor of 1 utilizes approximately 8 percent of the tape reel, leaving 92 percent of the reel unused—certainly not a very desirable situation. Clearly this large percentage of unused tape is a direct result of a small blocking factor.

Example 2. Let us use the same specifications defined in Example 1 with the exception of increasing the blocking factor from 1 to 20. This is represented by Fig. 8-4. We certainly hope to increase both the number of records on the medium and the efficiency of the medium. Figure 8-4 shows that a blocking factor of 20 provides 20 records in each 1.6 inches of tape. Therefore

30,000 inches/tape / 1.6 inches/block = 18,750 blocks/tape
and

18,750 blocks/tape × 20 records/block = 375,000 records/tape

Hence a blocking factor increased from 1 to 20 provides approximately 8 times more records on a tape medium. Correspondingly, the tape usage efficiency is improved since we now have

375,000 records/tape × .05 inches/record = 18,750 inches of
                                           records/tape

and

18,750 blocks/tape × .6 inches/block = 11,250.00 inches of
                                        IBG/tape

which reflects a tape consisting of approximately 63 percent data records. This is a dramatic improvement of efficiency.

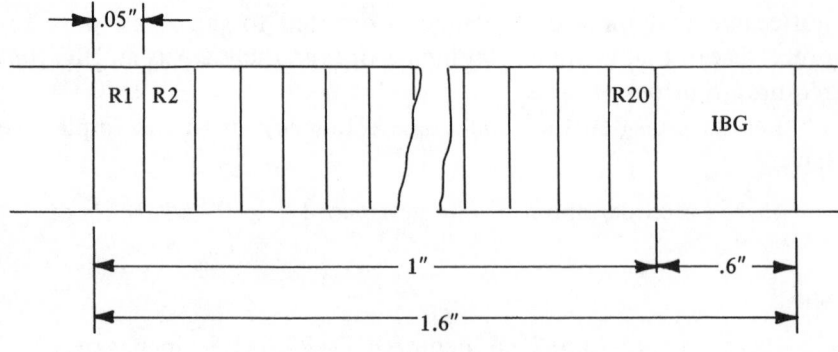

FIG. 8-4

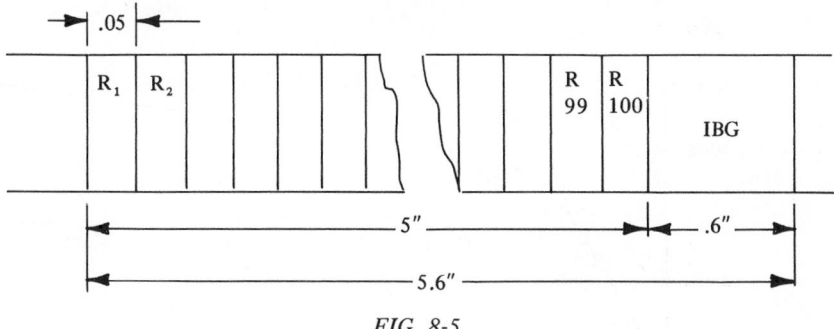

FIG. 8-5

Example 3. Let us consider one last example by increasing the blocking factor to 100, as illustrated in Fig. 8-5. A blocking factor of 100 means that 100 records will fit in 5.6 inches of tape. Therefore we see that

30,000 inches/tape / 5.6 inches/block = 5357 blocks/tape

and

5357 blocks/tape × 100 records/block = 535,700 records/tape

Furthermore, the efficiency is again increasing since

535,700 records/tape × .05 inches/record = 26,785 inches of
records/tape

and

5357 blocks/tape × .6 inches/block = 3214 inches of
IBG/tape

Now, the tape utilization efficiency is approaching 90 percent with the blocking factor increased to 100. This blocking factor permits us to store the equivalent of over one-half million punched cards on one reel of magnetic tape.

Finally, let us consider Figs. 8-6 and 8-7. Figure 8-6 is a partial graph that plots a function showing the relationship between the blocking factor and the number of records placed on the tape medium. Figure 8-7 is a partial graph that plots a function showing the relationship between the blocking factor and the efficiency percentage of the tape medium. Exercises 8-2 and 8-3 require the graphs to be expanded to include data for blocking factors of 40, 60, and 80, respectively.

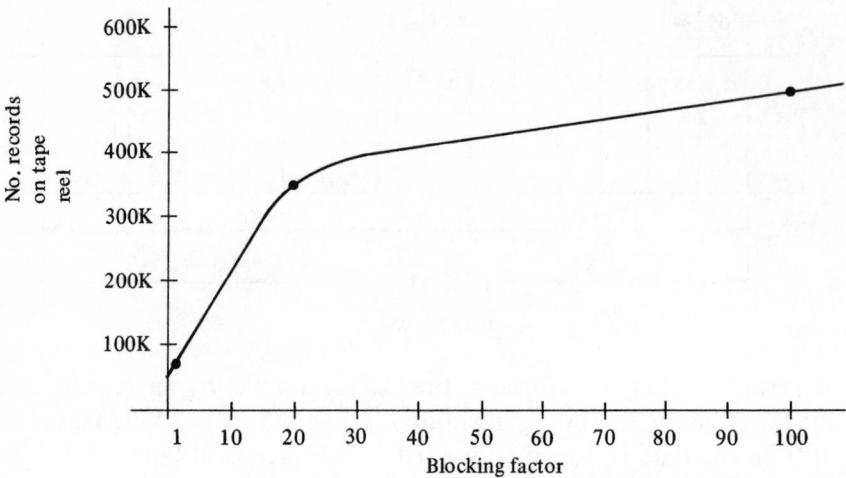

FIG. 8-6 Blocking factor vs. number records on medium.

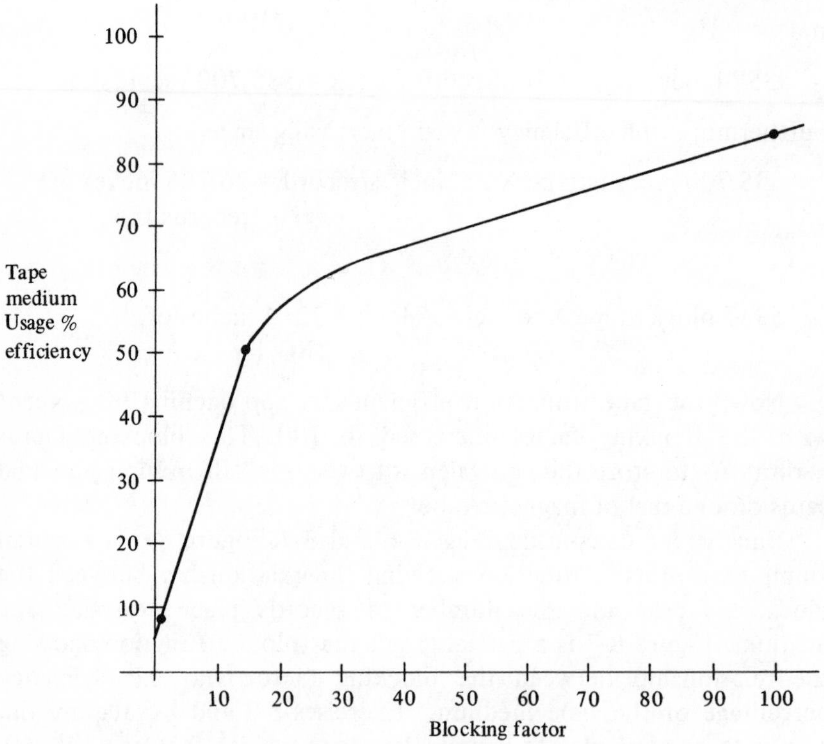

FIG. 8-7 Blocking factor vs. usage % efficiency.

EXERCISES

8-1. Given a master file of some 150,000 products, determine the number of tape reels required for the master file. Assume the following specifications: the tape reel is 2500 feet long; the tape density is 1600 CPI; the IBG is .6 inches long; the master records are 300 positions long; and the blocking factor is 5.

8-2. Using the specifications of Example 1 prepare a graph that plots the tape utilization percent versus the blocking factor. Use blocking factors of 1, 20, 40, 60, 80, and 100.

8-3. Prepare a graph that plots the tape record capacity versus the blocking factor. Use the blocking factors of 1, 20, 40, 60, 80, and 100.

To be sure, by increasing the blocking factor the tape utilization efficiency is increased. Therefore we may ask, How large may this blocking factor be? What constraints are to be considered in determining an appropriate blocking factor? To answer these questions, consider the basic elements of a computer program that consists of (1) the instructions (verbs) that manipulate and process data, such as READ, WRITE, MOVE, ADD, SUBTRACT, MULTIPLY, or DIVIDE, and (2) the data records that are processed by the above instructions. Figure 8-8 is a COBOL program that demonstrates this concept. Let us assume that the entire program is 2000 storage positions long. The data records are defined in the DATA DIVISION. The FILE SECTION defines an 80-position input record and a 133-position output record. The WORKING-STORAGE SECTION defines eight level 77 areas for a total of 34 positions and three level 01 records for a total of 399 positions. Hence the DATA DIVISION encompasses some 646 storage positions of records and work areas that will be processed by the 2000 minus 646 storage positions of instructions defined in the PROCEDURE DIVISION.

The union of the instruction set and data set comprises the program. Both of these subsets of a program require the use of a certain amount of main memory. The size of main memory is a constraint. Assume that a tape master record is unblocked and 500 positions long. A program will *read* that master file sequentially into a 500-position input location. If the master file is blocked 10, an entire set or block of ten records will be transferred into memory as a result of a read instruction. Therefore, a 5000-position input area

IBM

COBOL Coding Form

SYSTEM		PUNCHING INSTRUCTIONS		PAGE	OF
PROGRAM		GRAPHIC			*
PROGRAMMER	DATE	PUNCH	CARD FORM #		IDENTIFICATION

```
SEQUENCE  A  B
(PAGE)(SERIAL)        COBOL STATEMENT
01
02        DATA DIVISION.
03        FILE SECTION.
04        FD  PAYROLL-FILE
05            RECORDING MODE IS F
06            LABEL RECORDS ARE OMITTED.
07            DATA RECORD IS PAYROLL-RECORD.
08        Ø1  PAYROLL-RECORD.
09            Ø2  EMPL-NO    PICTURE X(9).
10            Ø2  EMPL-NAME  PICTURE X(2Ø).
11            Ø2  YTD-EARN   PICTURE 9(5)V99.
12            Ø2  YTD-FICA   PICTURE 9(4)V99.
13            Ø2  CURR-HRS   PICTURE 9(3)V99.
14            Ø2  RATE       PICTURE 99V99.
15            Ø2  DEPEND     PICTURE 99.
16            Ø2  FILLER     PICTURE X(24).
17
18        FD  REGISTER-FILE
19            RECORDING MODE IS F
20            LABEL RECORDS ARE OMITTED.
21            DATA RECORD IS PRINT-RECORD.
22        Ø1  PRINT-RECORD.
23            Ø2  FILLER     PICTURE X(133).
24
```

FIG. 8-8

IBM

COBOL Coding Form

SYSTEM			PUNCHING INSTRUCTIONS			PAGE	OF
PROGRAM		GRAPHIC			CARD FORM #	*	IDENTIFICATION
PROGRAMMER	DATE	PUNCH					73 [80

```
SEQUENCE  A   B                        COBOL STATEMENT
(PAGE) (SERIAL) CONT

01      WORKING-STORAGE SECTION.
02      77  PAGE-CT     PICTURE  999       VALUE  ZEROES.
03      77  LINE-CT     PICTURE  999       VALUE  ZEROES.
04      77  STD-PAY     PICTURE  999V99    VALUE  ZEROES.
05      77  OVT-PAY     PICTURE  999V99    VALUE  ZEROES.
06      77  GROSS-PAY   PICTURE  999V99    VALUE  ZEROES.
07      77  FICA        PICTURE  999V99    VALUE  ZEROES.
08      77  NET-PAY     PICTURE  999V99    VALUE  ZEROES.
09      Ø1  HEAD1.
10          Ø2  FILLER   PICTURE  X(12Ø)   VALUE  'WEEKLY PAYROLL REGISTER   PAGE'.
11          Ø2  PAGE-NO  PICTURE  ZZ.
12          -
13          -
14          Ø2  FILLER   PICTURE  X(111)   VALUE  SPACES.
15      -
16      Ø1  HEAD2.
17          Ø2  FILLER   PICTURE  X(13)    VALUE  'EMPL NUMB'.
18          Ø2  FICA
19          -   FILLER   PICTURE  X(12Ø)
20          Ø2                              VALUE  'HOURS   RATE   STAND   OVT
                                            YTD EARN   YTD   GROSS   FIT   FICA
                                            YTD FIT   NET   EMPLOYEE'.
```

FIG. 8-8 (cont.)

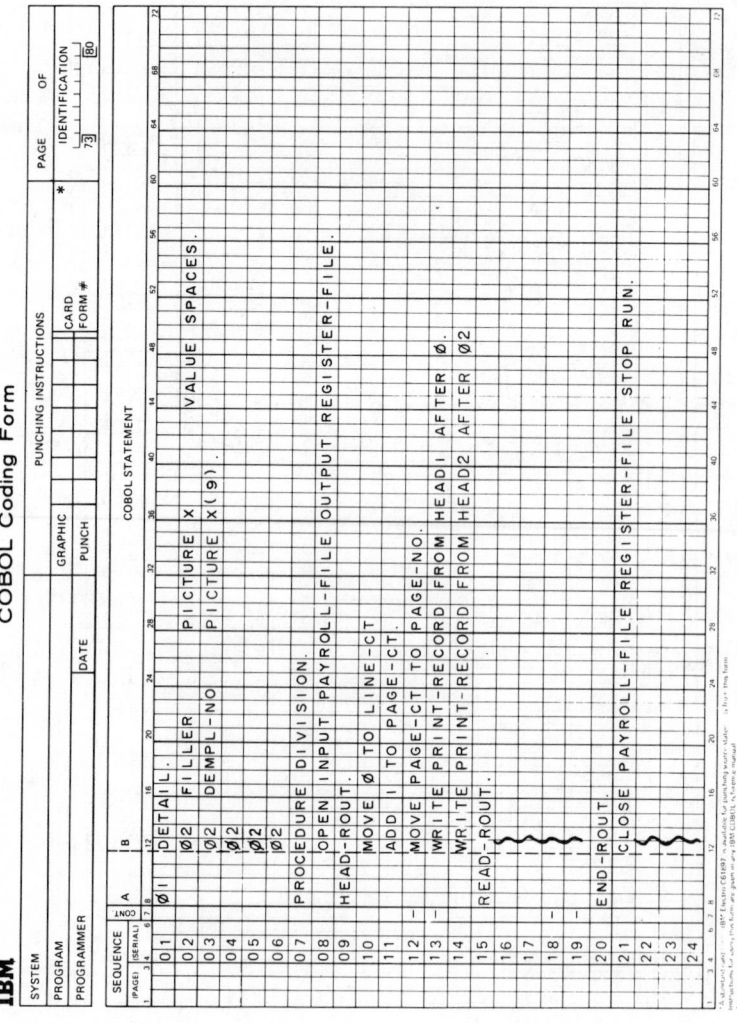

FIG. 8-8 (cont.)

must be reserved in main storage for that master file. And, if the master file is blocked 100, an unreal situation would exist where 50,000 storage positions would have to be reserved for the processing of the master file. This clearly would be, of course, impossible in most configurations.

Some small- and medium-scale computer configurations have small main storage capacities. Small capacities may range from 1000 (1K) to, say, 32,000 (32K) storage positions. Therefore, generally the programs should be developed to be able to fit into the available storage capacity. If a program has an instruction set that is, for example, 12,000 positions long, only 4000 positions may be reserved for the data set if main storage is 16K. Considering, for example, that there is only one master file to be processed and the records are 500 characters long it would be difficult to use too large a blocking factor for the master file. A blocking factor of 3 or 4 might be appropriate, depending on the number of other data elements and the amount of future growth anticipated in the program. These are considerations to be dealt with by the data processor.

Magnetic Disk

Magnetic disk is somewhat similar in concept to that of magnetic tape in that data are sorted along circular tracks using the same coding system as tape. The IBM 2311 disk storage drive (Fig. 8-9) requires the use of a removable disk pack (Fig. 8-10). This pack consists of six metal phonograph-record-like disks mounted on a vertical shaft. The disk drive has access arms that read and write on the removable disk pack. The upper surface of the top disk and bottom surface of the lower disk excluded, data are stored as magnetized bits in concentric tracks along each of the ten remaining recording surfaces. This removable disk pack, when inserted in the disk drive, makes up to over 7 million characters of data available to the system. The disk data surface, like the magnetic tape surface, can be used repetitively, with the new data being written on a track replacing the old data.

The disk pack may be thought of as consisting of two hundred concentric cylinders, Fig. 8-11, each of which has 10 tracks, with a capacity of approximately 3600 storage positions per track for a disk pack capacity of 7.2 million storage positions.

FIG. 8-9   IBM 2311 disk storage drive.

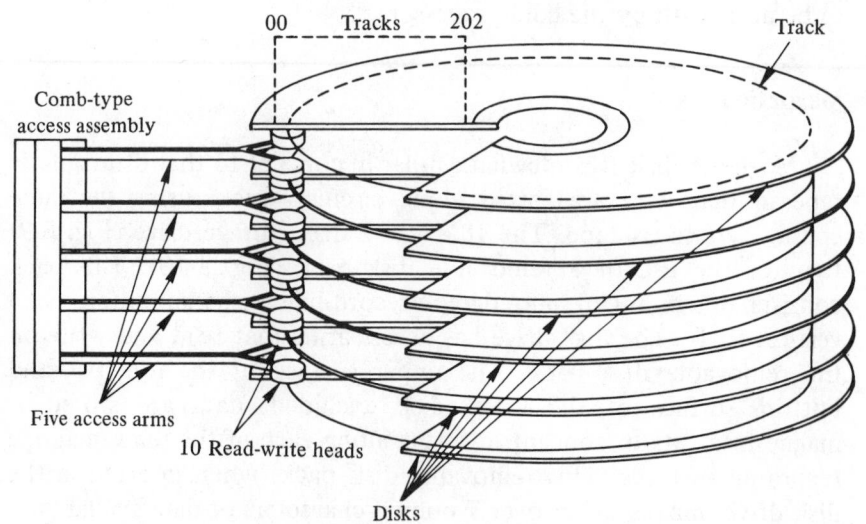

FIG. 8-10   IBM 2311 disk storage drive showing mounted disk pac and access arms.

Let us determine the record capacity of a disk pack using the record specifications given for the tape media previously in the chapter and compare the respective capacities of tape and disk.

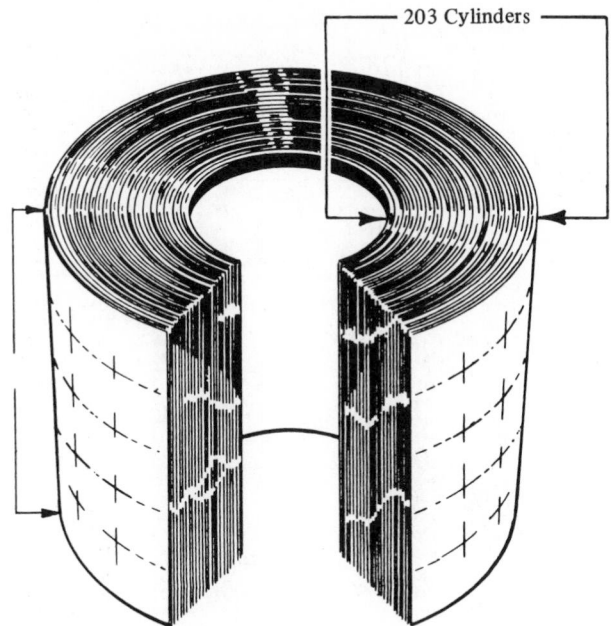

FIG. 8-11   IBM 2311 cylinders.

To determine the number of records that will reside on the disk media, the track capacity table shown in Fig. 8-12 can be a considerable aid to the data processor. In observing this table, it should be noted that there are two columns of information applicable to the 2311 device. One column is applicable to sequential file organization, and the other is applicable to random file organization such as indexed sequential. Indexed sequential file organization allows for the random retrieval of records of a file. This technique utilizes indexes, or tables, that indicate the key fields appearing on each track and cylinder. Keys generally used include such fields as product number, customer number, or employee number, for example.

Let us assume the records to be formatted without keys, as in a sequential disk application, and observe the column used for the 2311 device. The maximum number of characters that will be found on a track is 3625. Therefore, we could place one 3625 character record on a track, or if the record were 1800 characters long blocked 2, we could place 1 block or 3600 characters on a track.

Maximum Bytes per Record Formatted without Keys						*Records per Track*	*Maximum Bytes per Record Formatted with Keys*					
2311	2314	2302	2303	2301	2321		2311	2314	2302	2303	2301	2321
3625	7294	4984	4892	20483	2000	1	3605	7249	4964	4854	20430	1984
1740	3520	2403	2392	10175	935	2	1720	3476	2383	2354	10122	920
1131	2298	1570	1558	6739	592	3	1111	2254	1550	1520	6686	576
830	1693	1158	1142	5021	422	4	811	1649	1139	1104	4968	406
651	1332	912	892	3990	320	5	632	1288	893	854	3937	305
532	1092	749	725	3303	253	6	512	1049	730	687	3250	238
447	921	634	606	2812	205	7	428	877	614	568	2759	190
384	793	546	517	2444	169	8	364	750	527	479	2391	154
334	694	479	447	2157	142	9	315	650	460	409	2104	126
295	615	425	392	1928	119	10	275	571	406	354	1875	103
263	550	381	346	1741	101	11	244	506	362	308	1688	85
236	496	344	308	1585	86	12	217	452	325	270	1532	70
213	450	313	276	1452	73	13	194	407	294	238	1399	58
193	411	286	249	1339	62	14	174	368	267	211	1286	47
177	377	264	225	1241	53	15	158	333	245	187	1188	38
162	347	244	204	1155	44	16	143	304	224	166	1102	29
149	321	225	186	1079	37	17	130	277	206	148	1026	21
138	298	209	169	1012	30	18	119	254	190	131	959	15
127	276	196	155	952	24	19	108	233	176	117	899	9
118	258	183	142	897	20	20	99	215	163	104	844	
109	241	171	130	848	15	21	90	198	152	92	795	
102	226	161	119	804	10	22	82	183	142	81	751	
95	211	151	109	763	6	23	76	168	132	71	710	
88	199	143	100	726		24	69	156	123	62	673	
82	187	135	92	691		25	63	144	116	54	638	
77	176	127	84	659		26	58	133	108	46	606	
72	166	121	77	630		27	53	123	102	39	577	
67	157	114	70	603		28	48	114	95	32	550	
63	148	108	64	577		29	44	105	89	26	524	
59	139	102	58	554		30	40	96	83	20	501	

FIG. 8-12   Track capacity table.

In the first tape example, the records are 80 characters long and unblocked. Looking down the 2311 column to 82 and looking across to the number of records per track column, we find 25. Hence 25 unblocked, 80-character records can be placed on a track. This is 2000 characters, which means approximately 1600 positions per track remain unused. Therefore, we find that

25 records/track × 2000 tracks/pack = 50,000 records/pack

Notice that the 50,000 records somewhat parallel the tape capacity of 46,153 records, using the 80-character unblocked record example. Now, consider a blocking factor of 20. Hence a record is 20 times 80 or 1600 characters in length. Two of these blocks, or 40 records, can be placed on a track using 3600 of a possible 3625 storage positions. This is close to an optimum utilization of the track capacity, with the optimum blocking factor to be 45 records per block. Thus

40 records/tracks × 2000 tracks/pack = 80,000 records/pack

Now, it must be noticed that while we have efficiently used the disk pack, we have placed only 80,000 records on the medium. Eighty-thousand records are far less than the 375,000 records placed on the tape in the second example previously. However, the smaller numbers of records afforded a data processor by disk may be offset by file organization capabilities offered by disk and not available on tape. That is, a tape system allows only for the sequential accessing of records, whereas, with disk, we have the capability of randomly retrieving a data record no matter which of the 2000 tracks that record is located on. This capability is required in some EDP applications such as the airline reservation system or banking systems in which immediate response is required. Tape, on the other hand, lends itself to volatile applications such as inventory control or payroll systems where a large percentage of the master file will be accessed and updated.

EXERCISES

8-4. Assume that a payroll system is to be implemented. The master file consists of 20,000 records each 149 bytes long. Sequential file organization is to be used. The hardware consists of three 2311 magnetic disk drives, card reader punch, printer, and 32K CPU. How many disk packs will be required for a blocking factor of 10? What would the optimum blocking factor be?

8-5. Prepare graphs for the disk medium as were prepared for tape in Exercises 8-2 and 8-3.

Chapter 9

LINEAR PROGRAMMING

Inequalities

An inequality is an expression whose appearance and attributes are much like that of an equation. If differs from an equation in that it will have a greater than ($>$) or less than ($<$) sign in addition to, or instead of, an equal sign. Therefore

$$x > 3$$

$$5x \geqslant 4$$

$$4x < -3$$

$$2x + 4 < 6x - 7$$

$$5 < x$$

are examples of inequalities.

Just as in dealing with equations we are interested in those values of the variable that make the inequality true. For instance, in considering the inequality

$$x > 3$$

and replacing x by a negative number, 0, 1, 2, or 3, we see that

$$x \not> 3 \quad (x \text{ is not greater than } 3)$$

Solutions to this particular inequality will consist of those numbers greater than 3, as shown in Fig. 9-1.

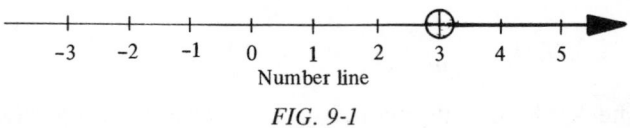

Number line

FIG. 9-1

Inequalities in one variable will be pictorially represented by the use of the number line, as in Fig. 9-1. Therefore numbers to the right of 3 make the above inequality true, whereas 3 and numbers to the left of 3 make the expression false. If the inequality were

$$x \geqslant 3$$

the number 3 would also be a valid solution. In Fig. 9-1 the small circle for the number 3 means that 3 is not a solution to the inequality.

As opposed to equations where we found one, two, or three solutions, an inequality may have many solutions. For example, there are an infinite number of solutions in the above expression.

In solving inequalities it is convenient to understand properties of inequalities. First, let us agree that two inequalities have the same sense if their signs of inequality point in the same direction. Therefore we have the following three properties:

1. The sense of an inequality is not changed if the same number is added to (or subtracted) from both sides of the inequality. That is, if

$$a > b$$

then

$$a + c > b + c$$

and

$$a - c > b - c$$

2. The sense of an inequality is not changed if both sides of the inequality are multiplied (or divided) by the same positive number.

That is, if

$$a > b \quad \text{and} \quad c > 0$$

then

$$ac > bc$$

3. The sense of an inequality is reversed if both sides of the inequality are multiplied (or divided) by the same negative number. That is, if

$$a > b \quad \text{and} \quad c < 0$$

then

$$ac < bc$$

Therefore given the inequality

$$5x + 3 < 2x - 8$$

it is necessary only to transpose +3 and $2x$ by adding

$$-2x - 3$$

to both sides of the inequality. Then, dividing both sides of the inequality by 3, we have

$$5x + 3 < 2x - 8$$
$$5x + 3 - 2x - 3 < 2x - 8 - 2x - 3$$
$$3x < -11$$

and

$$x < -\frac{11}{3}$$

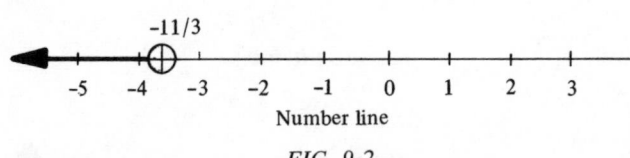

FIG. 9-2

This solution is demonstrated in Fig. 9-2. Considering

$$x = -3$$

as a possible solution we see that

$$5(-3) + 3 \overset{?}{<} 2(-3) - 8$$

$$-15 + 3 \overset{?}{<} (-6) - 8$$

$$-12 \not< -14$$

indicating that

$$x = -3$$

is not a solution. Considering

$$x = -4$$

we have

$$5(-4) + 3 \overset{?}{<} 2(-4) - 8$$

$$-20 + 3 \overset{?}{<} -8 - 8$$

$$-17 < -16$$

and

$$x = -4$$

is a solution.

Given the inequality

$$2x + 3 > 3x - 7$$

we find the solution to be

$$2x - 3x > -7 - 3$$

$$-x > -10$$

$$x < 10$$

Notice that by dividing both sides of the inequality by -1 the sense of the inequality is reversed. Figure 9-3 graphically represents the solution to the above inequality. We can test our solution by using values of x in the vicinity of 10. For example, letting

$$x = 9$$

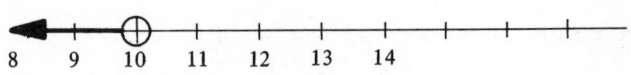

Number line

FIG. 9-3

we find that

$$2(9) + 3 \overset{?}{>} 3(9) - 7$$

$$18 + 3 \overset{?}{>} 27 - 7$$

$$21 > 20$$

Letting

$$x = 10$$

we find that

$$2(10) + 3 \overset{?}{>} 3(10) - 7$$

$$20 + 3 \overset{?}{>} 30 - 7$$

and

$$23 \not> 23$$

Therefore

$$x = 9$$

is a solution and

$$x = 10$$

is not a solution to the inequality.

Sometimes it is necessary to find solutions to two simultaneous inequalities. For example, given

$$x > 2 \quad \text{and} \quad x < 6$$

we see that x must be greater than 2 and less than 6, as shown in Fig. 9-4. The solution to

$$x > 2$$

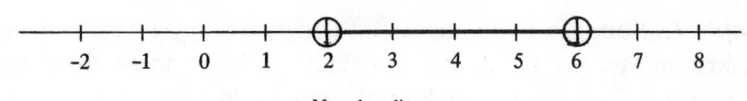

Number line

FIG. 9-4

is the set

$$S_1 = \{x; x > 2\}$$

The solution to

$$x < 6$$

is the set

$$S_2 = \{x; x < 6\}$$

The solution to the simultaneous inequalities is the set formed by the intersection

$$S_3 = S_1 \cap S_2$$

EXERCISE

9-1. Find solutions to the following inequalities:

(a) $2x > 7$

(b) $-2x \geqslant -7$

(c) $4x + 3 < 6x$

(d) $3 < 4x + 7$

(e) $5x + 10 > 15$ and $x < 3$

(f) $x < -5$ and $x < -7$

(g) $x < -5$ and $x > 5$

(h) $x \geqslant -5$ and $x \leqslant 5$

(i) $\frac{4}{3}y < \frac{2}{3}y + 1$

(j) $-\frac{2}{5}y > \frac{5}{2}$

(k) $\frac{3}{2} - y < \frac{2}{3}y + \frac{3}{5}$

Example 1. Suppose that a manufacturing plant produces two products, brooms and mops. A broom requires 2 hours to produce, and a mop requires 3 hours. How many of each could be produced in a 12-hour period? The inequality

$$2B + 3M \leqslant 12$$

satisfies this quesiton, with B and M representing the number of brooms and mops produced in the 12-hour period of time. Realizing that only integer values for B and M are possible let us consider the following solutions:

Number of Brooms	Number of Mops	Manufacturing Time (hours)
6	0	12
5	0	10
4	1	11
3	2	12
2	2	10
1	3	11
0	4	12

In a problem such as this we are searching for optimum solutions. Note that while we could produce other combinations of the products in a 12-hour period we are not interested in a solution such as, say, one mop and two brooms. The equation

$$2B + 3M = 12$$

is shown in Fig. 9-5. The shaded area represents all possible solutions to the inequality

$$2B + 3M \leqslant 12$$

that are defined within the problem constraints.

Point A, representing three brooms and two mops, is within the shaded area and is a possible solution. Point B, representing five brooms and three mops, is not within the shaded area and therefore is not a possible solution. Points on the graph of the equation

$$2B + 3M = 12$$

are considered as possible solutions.

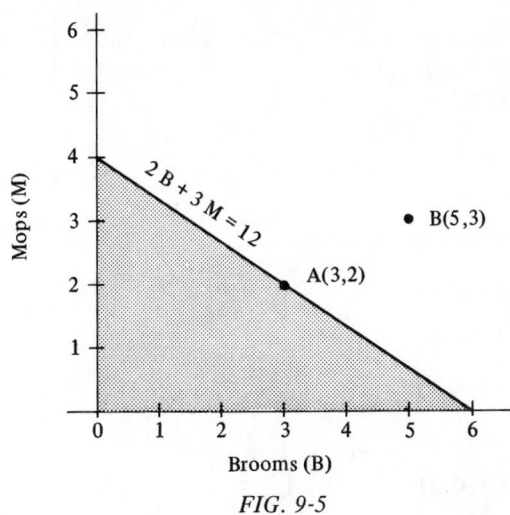

FIG. 9-5

Suppose that the profit on a broom is $2 and the profit on a mop is $3. We can now expand our optimum solution table to reflect this and show the associated profit for each combination. Hence

Number of Brooms	Number of Mops	Manufacturing Time (hours)	Profit ($)
6	0	12	12
5	0	10	10
4	1	11	11
3	2	12	12
2	2	10	10
1	3	11	11
0	4	12	12

indicating that the most profitable combinations of products to manufacturer are (6,0), (3,2), and (0,4). This is, in fact, one of the considerations involved in linear programming.

EXERCISES

9-2. Assume that in Example 1 a broom requires 1 hour to produce and a mop requires 2 hours. What would be the best mix of products for a 40-hour work week to optimize profits?

9-3. Prepare graphs that illustrate the area representing possible solutions for the following constraint equations:

(a) $2B + 3M \leqslant 16$

(b) $3B + 3M \leqslant 24$

(c) $2B + 3M \leqslant 16$

$3B + 3M \leqslant 24$

(d) $1.5X + 2.5Y \leqslant 20$

$2.0X + 3.5Y \leqslant 18$

(e) $1.5X + 2.5Y \leqslant 20$

$2.0X + 3.0Y \leqslant 16$

$3.0X + 3.0Y \leqslant 24$

Linear Programming

Linear programming encompasses mathematical techniques in finding optimum solutions to problems consisting of variables such as time or labor. The examples used in this section reflect only an intuitive introduction to linear programming.

Example 2. The Sweepclean Manufacturing Company produces brooms and mops by operations in two different work center (WC) locations. WC1 is in operation 320 hours per month while WC2 is in operation 240 hours per month. At WC1 it requires 1 hour to process a broom and 2 hours to process a mop, while at WC2 it takes 2 hours to process a broom and 1 hour to process a mop. The company makes a profit of $1 per broom and $2 per mop. Find the number of brooms and mops to be produced which will yield a maximum profit.

The equations representing the constraints for each WC are

$$1B + 2M \leqslant 320 \quad \text{(WC1)}$$

$$2B + 1M \leqslant 240 \quad \text{(WC2)}$$

where B and M represent the number of brooms and mops produced per month. Figure 9-6 depicts the two equations, with the shaded area representing the feasible area, or sets of values for B and M that satisfy both inequalities simultaneously.

The profit equation is

$$1B + 2M = P$$

which we are to maximize. This equation defines the profit to be equal to the number of brooms times \$1 plus the number of mops times \$2. We are searching for those values of B and M that produce the greatest profit and yet intersect our feasible region.

Let us assign a trial value to P in order to comprehend what the profit equation resembles. Letting

$$P = 160$$

the equation

$$1B + 2M = 160$$

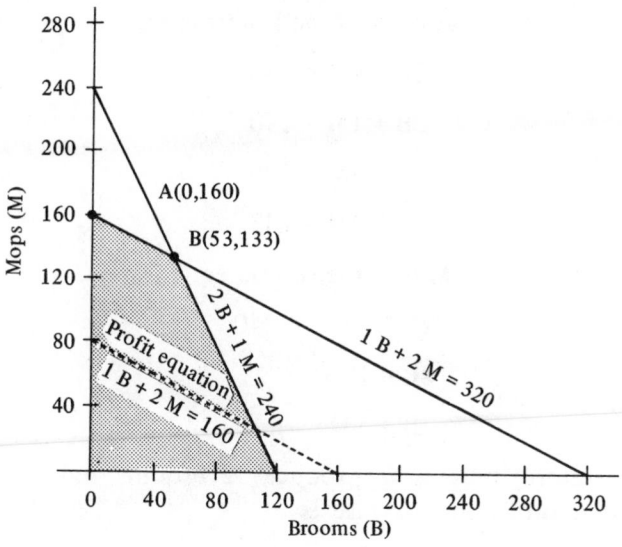

FIG. 9-6

is shown by the dotted curve of Fig. 9-6. The profit equation has the same slope as

$$1B + 2M = 320$$

which is our constraint for WC1. Therefore, our maximum values for B and M will coincide with this equation.

Let us consider points A and B as possible solutions. Point A determines that the number of brooms and mops to be produced is zero and 160 units, respectively. Therefore, substituting into our equations for each WC we have

$$1(0) + 2(160) \leqslant 320 \qquad \text{(WC1)}$$

$$2(0) + 1(160) \leqslant 240 \qquad \text{(WC2)}$$

with a profit for each WC of

$$1(0) + 2(160) = \$320$$

Point B is found by solving the equations

$$1B + 2M = 320$$

$$2B + 1M = 240$$

multiplying the first equation by 2, and subtracting

$$2B + 4M = 640$$

$$2B + 1M = 240$$

$$3M = 400$$

$$M = 133\tfrac{1}{3}$$

substituting the value of M into the second equation

$$2B + 133\tfrac{1}{3} = 240$$

$$2B = 106\tfrac{2}{3}$$

$$B = 53\tfrac{1}{3}$$

Point B, therefore, tells us to produce 53 brooms and 133 mops. The associated profit in this case is

$$1(53) + 2(133) = \$319$$

which is nearly the same as the profit for point A. Furthermore, the required production time at the WCs is

$$1(53) + 2(133) = 319 \leqslant 320 \qquad \text{(WC1)}$$

and

$$2(53) + 1(133) = 239 \leqslant 240 \qquad (WC2)$$

Now it is a question of choosing one of the combinations defined by points A and B. Since point A requires the termination of production of one of the products, let us choose point B as the solution.

Example 3. Let us modify the parameters of Example 2. Assume that the hours constraint equations are

$$1B + 2M \leqslant 320 \qquad (WC1)$$

$$1B + 1M \leqslant 200 \qquad (WC2)$$

with a profit equation of

$$2B + 3M = P$$

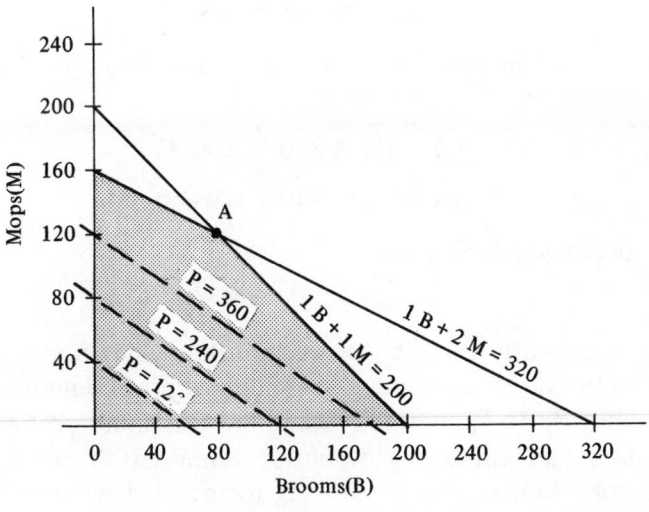

FIG. 9-7

These equations are shown in Fig. 9-7 with the shaded area representing the feasible area and the dotted lines representing the family of profit equations. It can be observed that the family of profit curves will intersect point A, with point A, therefore, representing optimum values for B and M. Subtracting the constraint equations, we have

$$M = 120$$

Therefore

$$B = 80$$

Hence 80 brooms and 120 mops should be manufactured. The inequalities are satisfied, since

$$1(80) + 2(120) = 320 \leqslant 320 \quad \text{(WC1)}$$

$$1(80) + 1(120) = 200 \leqslant 200 \quad \text{(WC2)}$$

and the profit is

$$2(80) + 3(120) = \$520$$

Compare this to the profit if point B were used. Using 0 brooms and 160 mops the profit is

$$2(0) + 3(160) = \$480$$

Example 4. Let us modify the parameters of Example 2 one more time. Assume that the constraints are still

$$1B + 2M \leqslant 320 \quad \text{(WC1)}$$

$$2B + 1M \leqslant 240 \quad \text{(WC2)}$$

but the profit equation is now

$$.2B + .2M = P$$

That is, the profit is $.20 for each broom and mop. Figure 9-8 illustrates the three equations. The family of profit equations are the dotted lines P_1 to P_6, with P_6 being the maximum profit equation intersecting the feasible area at point A. Point A is the intersection of the WC constraint equations. Solving for point A we have the point

$$B = 53, \quad M = 133$$

as solved in Example 2. The optimum profit is therefore

$$P = .2(53) + .2(133)$$

$$= .2(53 + 133)$$

$$= .2(186)$$

$$= \$37.20$$

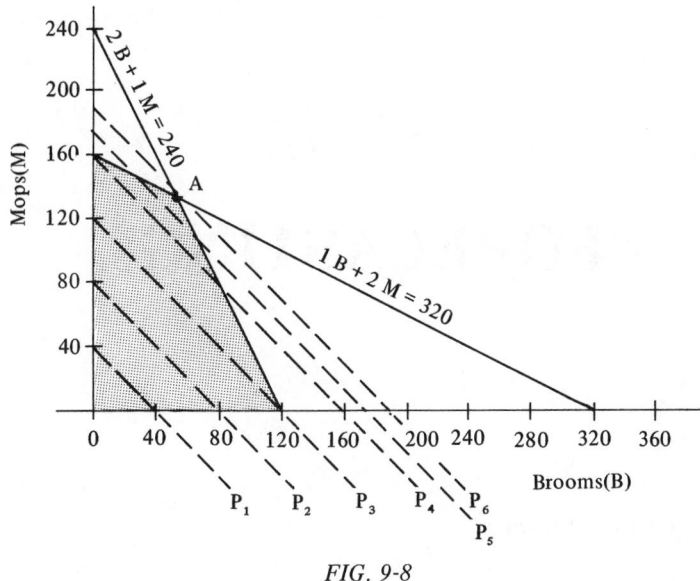

FIG. 9-8

EXERCISES

9-4. The Sweepclean Company has come into new management. WC1 and WC2 are now in operation 360 and 320 hours per month, respectively. WC1 now requires 1 hour to process a broom and 1.5 hours to process a mop, while WC2 requires 1 hour to process a broom or mop. The company makes $.30 on each broom and $.20 on each mop. Find the proper mix of products that will yield a maximum profit.

9-5. The XYZ Company makes products P and Q. P requires 2 hours of processing at WC1 and 1 hour at WC2. Q requires 1 hour at WC1 and 2 hours at WC2. WC1 and WC2 are in operation 104 and 76 hours per month respectively. P and Q return profits of $5 and $10 respectively. How many of each product should be produced each month to maximize profits?

Chapter 10

FORECASTING

Linear Equation Method

The business community is very concerned with meeting the demands for goods and products by their customers. Inventory levels must be maintained at levels that ensure the fulfilling of customer orders. If a customer goes to distributor ABC for an item and it is not in stock, the chances are excellent that the customer will go to distributor XYZ. Customer demand certainly may be satisfied by maintaining high inventory levels but this means that a great deal of money is tied up in excess inventory. Therefore a problem exists. It is necessary to develop inventory levels that effectively satisfy a very high percentage of customer demand while yet minimizing the inventory levels.

This problem is solved by developing effective methods for forecasting future customer demand. Forecasting encompasses the mathematical manipulation of historical data. It should be noted that the techniques discussed in this chapter are very basic and merely serve to introduce the data processing student to forecasting.

In Chapter 1, Examples 10 and 11 dealt with the fitting of a linear equation (straight line) to a set of points, with the points representing the sales history of a product for past periods of time. The data in Example 10 were chosen so that all the points coincided with a particular equation. The data in Example 11, however, were chosen to generate a little difficulty in that they did not all coincide with a straight line. It was therefore necessary to find an equation that would approximately serve to represent all the data equally.

The sales history in Example 10 followed the sequence

$$10, 15, 20, \ldots, 50$$

for the first nine periods and the problem was to determine a forecast for the last three periods. A linear equation was generated using the two-point formula. Another method for finding the same equation available. The equation we are looking for is of the form

$$y = ax + b$$

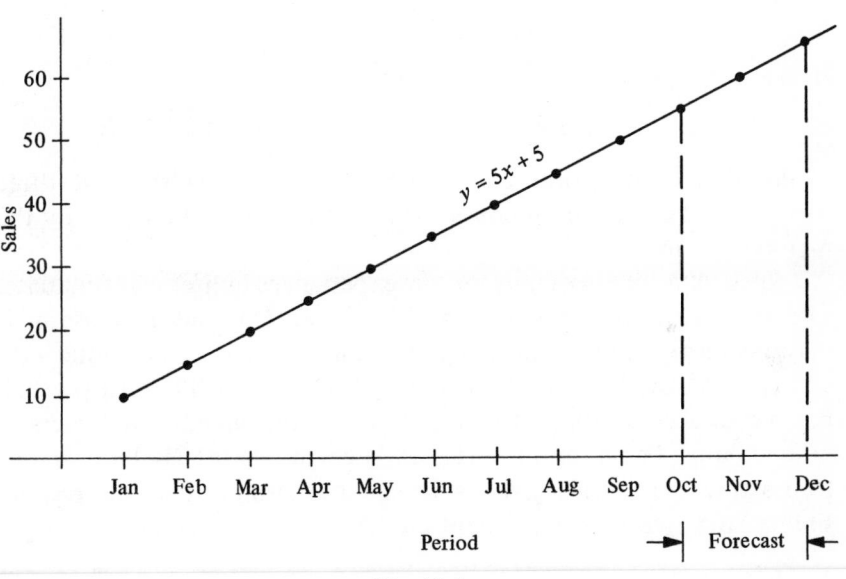

Fig. 10-1

as shown in Fig. 10-1. Therefore, given two points (1,10), and (2,15), where (x,y) is an ordered pair in which x, the first component, represents the period, and y, the second component, represents the volume, we can generate the following equations:

$$10 = a + b$$

$$15 = 2a + b$$

which can be solved by the use of determinants with the parameters a and b acting as the unknowns. Therefore

$$a = \frac{\begin{vmatrix} 10 & 1 \\ 15 & 1 \end{vmatrix}}{\begin{vmatrix} 1 & 1 \\ 2 & 1 \end{vmatrix}} = \frac{10 - 15}{1 - 2} = 5$$

and

$$b = \frac{\begin{vmatrix} 1 & 10 \\ 2 & 15 \end{vmatrix}}{\begin{vmatrix} 1 & 1 \\ 2 & 1 \end{vmatrix}} = \frac{15 - 20}{1 - 2} = 5$$

Hence

$$y = 5x + 5$$

is the desired equation to be used in forecasting the last three periods. This equation, of course, is identical to the line found by the method of Chapter 1.

Sales data, like so many of our experiences in life, will fluctuate. The volume, as forecasted by the last equation, may increase or it may decrease. There is no magic formula (model) to be found that will always exactly predict what the future sales will be. All we shall be able to accomplish is the capacity of making an educated guess of sales volume based upon variables such as historical data. If our forecasting techniques give us inaccurate results, it is our responsibility to derive a more efficient model.

Averaging Methods

One method that may be used in the *moving average* method of forecasting. Consider the bar graph of Fig. 10-2, which shows the sales history of a product for the first 6 months of a year. Certainly, it would be difficult to use a linear equation to forecast future demand. The sales performance in the first 6 months is 30, 40, 25, 45, 60, and 40 units, respectively. The forecast for the seventh

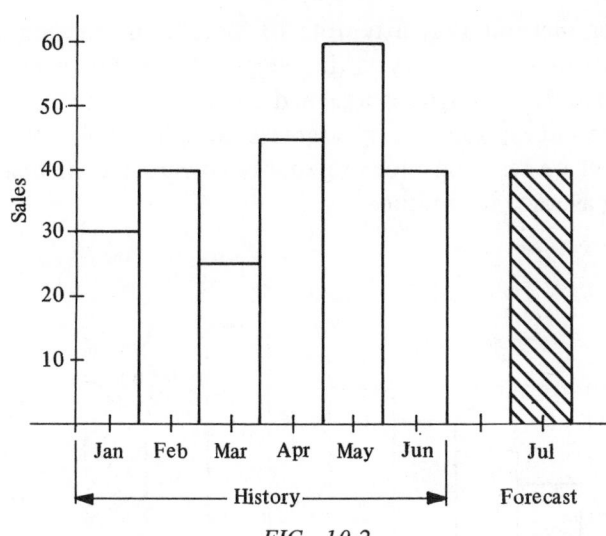

FIG. 10-2

month using the moving average method is 40 units. This average is found by solving the formula

$$\text{MOVING AVERAGE} = \Sigma \ \frac{\text{Units per period}}{\text{No. of periods}}$$

$$= \frac{30 + 40 + 25 + 45 + 60 + 40}{6}$$

$$= \frac{240}{6}$$

and

$$\text{MOVING AVERAGE} = 40$$

This method has some drawbacks. The main objection is that each period's demand has equal weight. This means that the demand from an earlier period is considered in the same manner as the demand of a more recent period. This is especially a weakness if the demand has not been constant but, in fact, defines a trend. The trend of the volume may be up or down, but, nevertheless, the most recent sales history conveys more meaningful data than the very early history.

Another method that attempts to place more emphasis on the later periods of sales history is the weighted moving average (WMA) method. That is, a weight is assigned to each historical period with the most recent period having a greater weight. Hence the forecast will in effect be reacting more favorably toward recent history than the moving average technique.

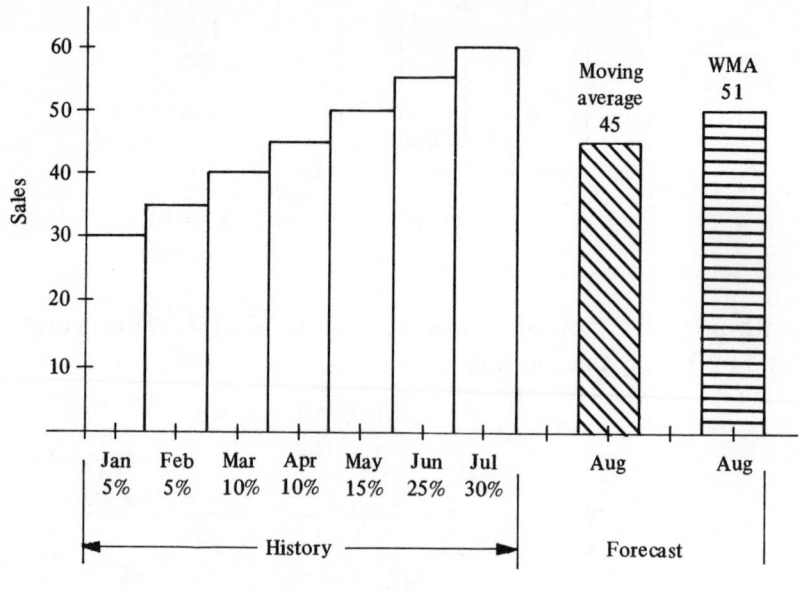

FIG. 10-3

Figure 10-3 contrasts the difference in the two methods of averaging. The sales for the first seven periods are 30, 35, 40, 45, 50, 55, and 60 units, respectively. Using the moving average method,

$$\text{MOVING AVERAGE} = \frac{30 + 35 + 40 + 45 + 50 + 55 + 60}{7}$$

$$= \frac{315}{7}$$

and

$$\text{MOVING AVERAGE} = 45$$

we arrive at a forecast of 45 units for the month of August. Obviously this anticipated usage does not take into account that the sales in each succeeding period is increasing. In other words, an upward trend is developing.

Using the weighted moving average technique and assigning more credence to the latter periods of history would seem to allow us to arrive at a more meaningful forecast. Let the weight be assigned as shown in Fig. 10-3, with the last three periods having the weights of 15, 25, and 30 percent, respectively. The formula for the weighted average is

$$WMA = \Sigma(Usage/Period \times Weight/Period)$$

Therefore

$$WMA = .05(30) + .05(35) + .1(40) + .1(45) + .15(50)$$
$$+ .25(55) + .3(60)$$

$$= .05(30 + 35) + .1(40 + 45) + .15(50) + .25(55) + .3(60)$$

$$= .05(65) + .1(85) + .15(50) + .25(55) + .3(60)$$

$$= 3.25 + 8.50 + 7.50 + 13.75 + 18.00$$

and

$$WMA = 51$$

The WMA should have been predicted to be greater than the moving average because of the greater emphasis on the recent periods. Based upon the sales history in this example the weighted averaging technique is to be preferred.

Let us see how the two methods compare when a decreasing sales trend has developed. This is shown in Fig. 10-4, which shows the forecast for August to be 45 units using the moving average method and 39 units using the weighted method. It should be clear that with the downward trend in progress the 39 is a more realistic forecast. We should also be aware that the weights assigned to each period are clearly variables that can be modified if future sales indicate it necessary.

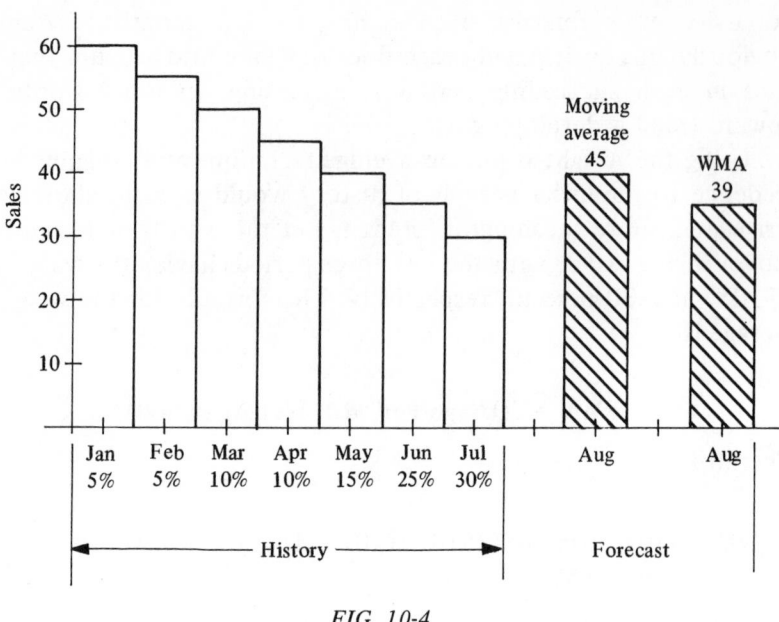

FIG. 10-4

EXERCISES

10-1. 20, 30, 40, 30, and 20 units have been the sales volume in the last five periods. What are the forecasts for the next period using the moving average method? Would the weighted moving average method be applicable, and if so, what weights would be assigned to each period and what is the WMA?

10-2. 40, 58, 82, 104, 130, and 148 units have been the sales volume for the first 6 months of the year. Discuss the development of a forecast for month 7.

We might take a moment to reflect on the three methods discussed so far. Before we are able to use a linear equation to represent historical data, the data must approximately coincide with a straight line. The data in Figs. 10-1, 10-3, and 10-4 are suitable for a linear equation; however, the data in Figs. 10-2, 10-5, and 10-6 are not suitable for this type of technique. Figure 10-2 is an illustration

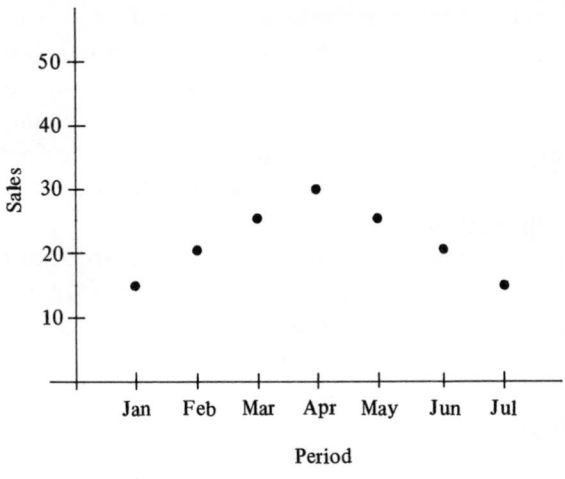

FIG. 10-5

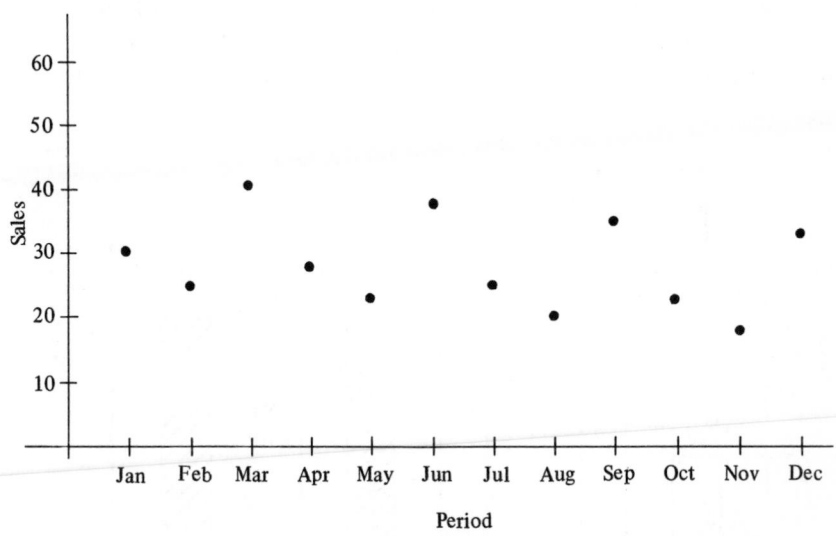

FIG. 10-6

of random data. No type of pattern can be observed. Figure 10-5 shows sales increasing linearly from January through April and then decreasing linearly the last three periods. Figure 10-6 shows the sales fluctuating in a pattern during each quarter of the year.

If the data approximately coincides with a linear equation, determine the equation for forecasting. Bear in mind, however, that there will be a point in time where the sales will deviate from the equation and a new forecasting method will have to be implemented.

If the data are random and do not indicate some form of trend that would allow us to give more weight to recent history, for instance, use the moving average method.

However, if a trend is indicated and the data are such that we are not able to use a linear equation, consider the WMA method with emphasis given to recent time periods.

Adjusting Weights

Consider the data represented by the bar graph in Fig. 10-7. We want to develop a weighted moving average for August. There would appear to be an upward trend in the historical data, and therefore we are going to weight the last three periods heavily. The weights, as shown in Fig. 10-7, are 10, 10, 20, 30, and 30 percent for the last

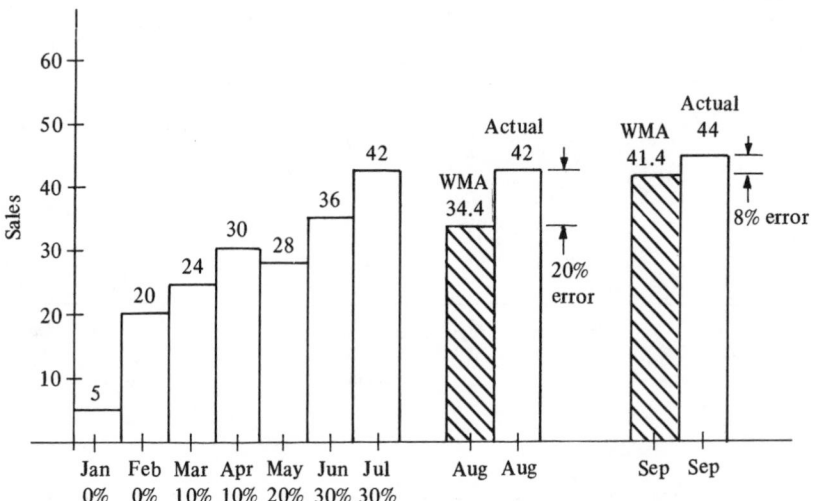

FIG. 10-7

five periods. The weighted moving average is found by solving

$$WMA = .1(24) + .1(30) + .2(28) + .3(36) + .3(42)$$
$$= 2.4 + 3.0 + 5.6 + 10.8 + 12.6$$

and

$$WMA = 34.4$$

We had calculated the forecast to be 34.4 units for August. The actual volume was 42 units. We have been approximately 20 percent in error. Assume that we desire our forecasts to remain within 10 percent of the actual volume. Hence it is going to be necessary to modify the weights that we have assigned to the historical periods. Let us reassign the weights to be 10, 20, and 70 percent of the last three months respectively. The WMA for August is now calculated to be

$$WMA = .1(28) + .2(36) + .7(42)$$
$$= 2.8 + 7.2 + 29.4$$

and

$$WMA(Aug) = 39.4 \text{ units}$$

which is approximately 8 percent in error. Furthermore, looking at the forecast for the next period we find that

$$WMA(Sep) = .1(36) + .2(42) + .7(42)$$
$$= 3.6 + 8.4 + 29.4$$

and

$$WMA(Sep) = 41.4 \text{ units}$$

and we are only approximately 6 percent in error of the actual sales of 44 units and within the desirable limit of this example.

Notice that by assigning weights to periods it is necessary that the summation of weights be equal to 1.0 (100 percent). For example, if we were going to base the WMA strictly on the volume of the previous period, we would multiply that volume by 1. Or, if we were going to base the WMA equally on the volume of the two previous periods, we would multiply each of the volumes by .5.

EXERCISE

10-3. Assume that the historical data are represented by the graph of Fig. 10-7 with the exception of September, which we have forecasted to be 41.4 using the weights of 10, 20, and 70 percent for the last three periods only. The actual sales for September dipped to 28 units, which is well outside our tolerance. What new formula will you develop for the forecasting of the next period?

Curve Fitting (Linear Regression)

Linear regression is a statistical technique that fits a linear equation, of the form $y = a + bx$, to a set of data points. This is a technique that minimizes the sum of the squares of the vertical deviations from the points to the line. The formulas for coefficients a and b are

$$a = \frac{\Sigma y \, \Sigma x^2 - \Sigma x \, \Sigma xy}{n \, \Sigma x^2 - (\Sigma x)^2}$$

$$b = \frac{n \, \Sigma xy - \Sigma x \, \Sigma y}{n \, \Sigma x^2 - (\Sigma x)^2}$$

where

$$\Sigma x = x_1 + x_2 + x_3 + \cdots + x_n$$

$$\Sigma y = y_1 + y_2 + y_3 + \cdots + y_n$$

$$\Sigma x^2 = x_1^2 + x_2^2 + x_3^2 + \cdots + x_n^2$$

$$\Sigma xy = x_1 y_1 + x_2 y_2 + x_3 y_3 + \cdots + x_n y_n$$

$$(\Sigma x)^2 = (x_1 + x_2 + x_3 + \cdots + x_n)^2$$

n = number of data points

For example, let us now reconsider the data from Example 5 of Chapter 1. The volume for January through September was 10, 13, 25, 25, 40, 15, 45, 55, and 60 units, respectively. Therefore

$$\Sigma x = 1 + 2 + 3 + 4 + 5 + 6 + 7 + 8 + 9$$

$$= 45$$

$$\Sigma\, y = 10 + 13 + 25 + 25 + 40 + 15 + 45 + 55 + 60$$
$$= 288$$

$$\Sigma\, x^2 = 1^2 + 2^2 + 3^2 + 4^2 + 5^2 + 6^2 + 7^2 + 8^2 + 9^2$$
$$= 1 + 4 + 9 + 16 + 25 + 36 + 49 + 64 + 81$$
$$= 285$$

$$\Sigma\, xy = (1 \cdot 10) + (2 \cdot 13) + (3 \cdot 25) + (4 \cdot 25) + (5 \cdot 40)$$
$$+ (6 \cdot 15) + (7 \cdot 45) + (8 \cdot 55) + (9 \cdot 60)$$
$$= 10 + 26 + 75 + 100 + 200 + 90 + 315 + 440 + 540$$
$$= 1796$$

$$(\Sigma\, x)^2 = 45^2$$
$$= 2025$$

$$n = 9$$

Hence

$$a = \frac{288 \cdot 285 - 45 \cdot 1796}{9 \cdot 285 - 2025}$$

$$= \frac{82{,}080 - 80{,}820}{2565 - 2025}$$

$$= \frac{1260}{540}$$

$$= 2.33$$

$$b = \frac{9 \cdot 1796 - 45 \cdot 288}{9 \cdot 285 - 2025}$$

$$= \frac{16{,}164 - 12{,}960}{2565 - 2025}$$

$$= \frac{3204}{540}$$

$$= 5.93$$

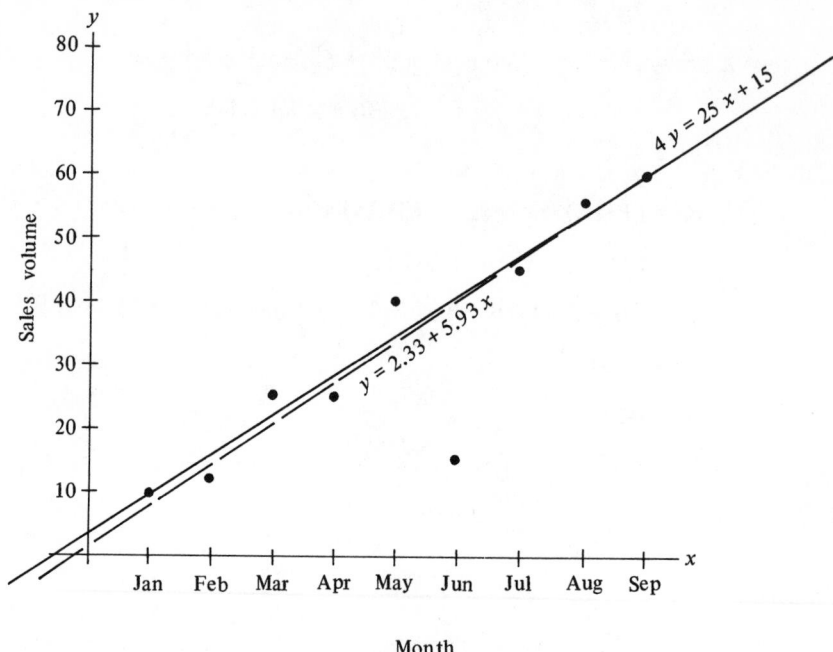

FIG. 10-8

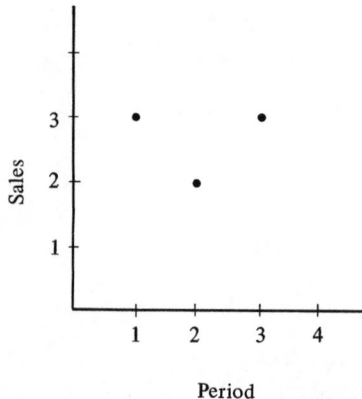

FIG. 10-9

Therefore our desired equation is

$$y = 2.33 + 5.93x$$

The equation is shown as a dotted line in Fig. 10-8 along with the equation derived in the original example. We are fortunate in this example since the two curves almost coincide.

Nonlinear Equations

Sometimes it may be advisable to find a higher degree equation to represent a set of historical data. This is illustrated by Fig. 10-9. For three periods we are given the volume of three, two, and three units. We are going to find an equation of the form

$$y = a_0 + a_1 x + a_2 x^2$$

which is called a second-degree polynomial or a quadratic equation. We have three sets of points, (1,3), (2,2), and (3,3), that must solve the desired equation. Therefore we generate the following three equations and three unknowns by substituting each of the sets of points into y:

$$3 = a_0 + a_1 + a_2$$

$$2 = a_0 + 2a_1 + 4a_2$$

$$3 = a_0 + 3a_1 + 9a_2$$

Three equations with three unknowns may be most readily solved by using determinants. We used determinants in Chapter 1 when solving two equations in two unknowns. Three equations, as shown above,

$$y_1 = a_0 + a_1 x_1 + a_2 x_1^2$$

$$y_2 = a_0 + a_1 x_2 + a_2 x_2^2$$

$$y_3 = a_0 + a_1 x_3 + a_2 x_3^2$$

are defined by substituting the three given points (x_1, y_1), (x_2, y_2), and (x_3, y_3) into the general equation

$$y = a_0 + a_1 x + a_2 x^2$$

where (x_1, y_1) is an ordered pair in which x_1 represents the period and y_1 represents the volume of the first period. The symbol

$$\begin{vmatrix} a_1 & b_1 & c_1 \\ a_2 & b_2 & c_2 \\ a_3 & b_3 & c_3 \end{vmatrix}$$

is called a determinant of third order whose value is defined as

$$\begin{vmatrix} a_1 & b_1 & c_1 \\ a_2 & b_2 & c_2 \\ a_3 & b_3 & c_3 \end{vmatrix} = a_1 b_2 c_3 + a_2 b_3 c_1 + a_3 b_1 c_2 - a_3 b_2 c_1 - a_1 b_3 c_2 - a_2 b_1 c_3$$

Hence

$$a_0 = \frac{\begin{vmatrix} 3 & 1 & 1 \\ 2 & 2 & 4 \\ 3 & 3 & 9 \end{vmatrix}}{\begin{vmatrix} 1 & 1 & 1 \\ 1 & 2 & 4 \\ 1 & 3 & 9 \end{vmatrix}} = \frac{54 + 6 + 12 - 6 - 36 - 18}{18 + 3 + 4 - 2 - 12 - 9}$$

$$= \frac{72 - 60}{25 - 23}$$

$$= \frac{12}{2} = 6$$

$$a_1 = \frac{\begin{vmatrix} 1 & 3 & 1 \\ 1 & 2 & 4 \\ 1 & 3 & 9 \end{vmatrix}}{\begin{vmatrix} 1 & 1 & 1 \\ 1 & 2 & 4 \\ 1 & 3 & 9 \end{vmatrix}} = \frac{18 + 12 + 3 - 2 - 12 - 27}{2}$$

$$= \frac{33 - 41}{2}$$

$$= \frac{-8}{2} = -4$$

$$a_2 = \frac{\begin{vmatrix} 1 & 1 & 3 \\ 1 & 2 & 2 \\ 1 & 3 & 3 \end{vmatrix}}{\begin{vmatrix} 1 & 1 & 1 \\ 1 & 2 & 4 \\ 1 & 3 & 9 \end{vmatrix}} = \frac{6 + 9 + 2 - 6 - 6 - 3}{2}$$

$$= \frac{17 - 15}{2}$$

$$= \frac{2}{2} = 1$$

Hence we find the second-degree equation to be

$$y = 6 - 4x + x^2$$

which, of course, does satisfy the data (1,3), (2,3), and (3,3). For period 4, we find the forecasted volume to be

$$\text{Volume} = 6 - 4(4) + 4^2$$

$$= 6 - 16 + 16$$

$$= 6 \text{ units}$$

The graph is shown in Fig. 10-10. If the three points that are being used to develop the quadratic equation coincide on a straight line, the equation will degenerate into a linear equation. In other words, if the volume for three consecutive periods were 10, 10, and 10, the points used to generate the second-degree equation would be (1,10), (2,10), and (3,10), respectively. The resulting equation would not be second-degree but would be the linear equation

$$y = 10 \text{ units}$$

The next example serves to illustrate sales that are cyclical. Let the sales for the first ten periods follow the sequence

$$5, 6, 5, 4, 5, 6, 5, 4, 5, 6$$

as shown in Fig. 10-11. The sales volume follows a pattern that is very much like a trigonometric function. General forms of a trigonometric function include

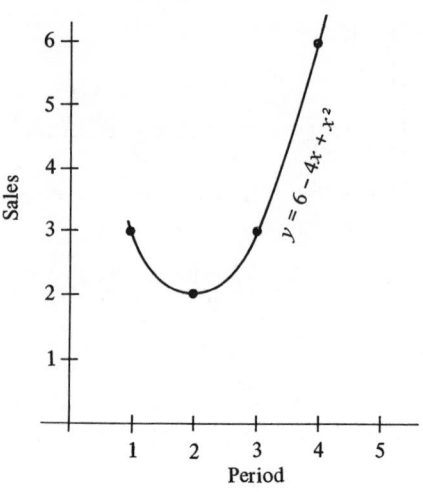

FIG. 10-10

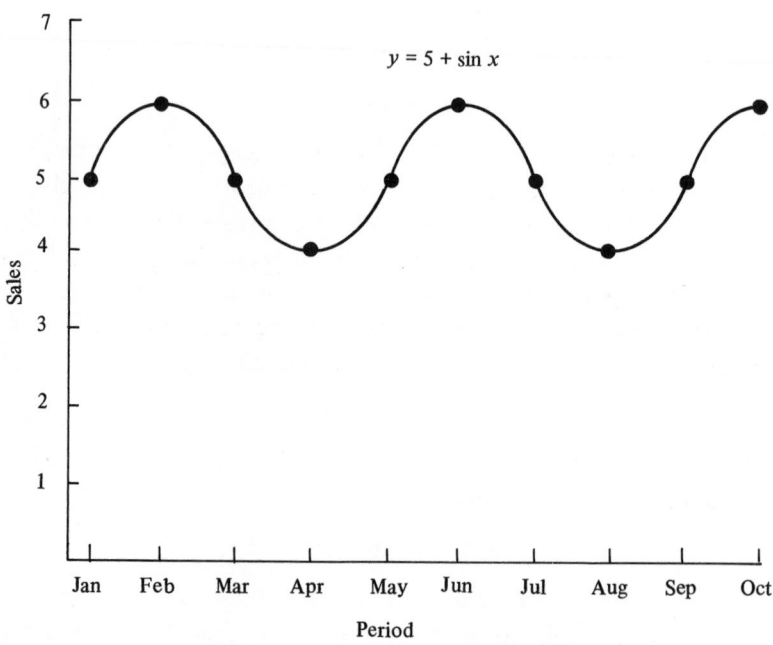

FIG. 10-11

$$y = a + b \sin x$$

$$r = c + d \cos t$$

In this case the data have been developed to follow the function

$$y = 5 + \sin x$$

EXERCISES

10-4. The volume for periods January through April for product XYZ has been 30, 25, 20, and 25 units, respectively. Determine the moving average and WMA for the May period. Find a second-degree equation that will pass through the last three periods and use this equation to forecast the volume for May.

10-5. The actual volume for May in Exercise 10-4 is 22 units. What forecasting method provided us with the best forecast? Do the weights that you used in the WMA method have to be modified because of the actual volume for May?

10-6. The volume for six consecutive periods was 9, 12, 22, 20, 24, and 28 units, respectively. Using the technique of linear regression find a linear equation that best fits the data.

10-7. The volume for January through August was 21, 24, 27, 30, 33, 36, 39, and 42 units, respectively. Using two methods find a linear equation that represents the data.

Chapter 11

SEARCHING LISTS

In data processing, it is often necessary to know the position that a value has in a list of values so that the position can be used to locate a corresponding value in *another* list. For example, suppose that we have two lists, each list containing 100 values. The beginning values of those lists are shown in Fig. 11-1. As the list names indicate, one of the lists contains pay numbers organized in increasing sequence. The other list contains corresponding pay rates. Thus the pay rate of the worker whose pay number is 12463 is $4.85; the pay rate of the person whose pay number is 12489 is $3.76; etc.

PAY-NUM		PAY-RATE	
12463		$4.85	
12489		$3.76	
12554		$5.14	
12593		$3.80	
12603		$4.73	
12613		$4.25	
12620		$3.66	

Etc.   Etc.

FIG. 11-1

Assume that the computer is given a pay number that it is to find in the pay number list and then give the corresponding pay rate. The computer executes a set of instructions (a program) to find the desired information. That set of instructions is written so that it does not matter which pay number is involved; the program will be able to find it.

The flowchart in Fig. 11-2 shows how the program finds the desired pay number and prints the corresponding value of pay rate. To understand the flowchart, begin at the circle labeled A. If you follow the arrows, you will see what instructions must be given to the computer and in what sequence.

The shapes of the symbols in the flowchart are important. They allow an easier understanding of what is to be done. For instance, the following symbol (parallelogram), as noted previously, is used for I/O (input/output).

The symbol

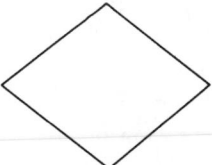

indicates that a decision must be made by the program. Details of the decision's nature are written within the diamond.

The symbol

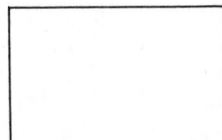

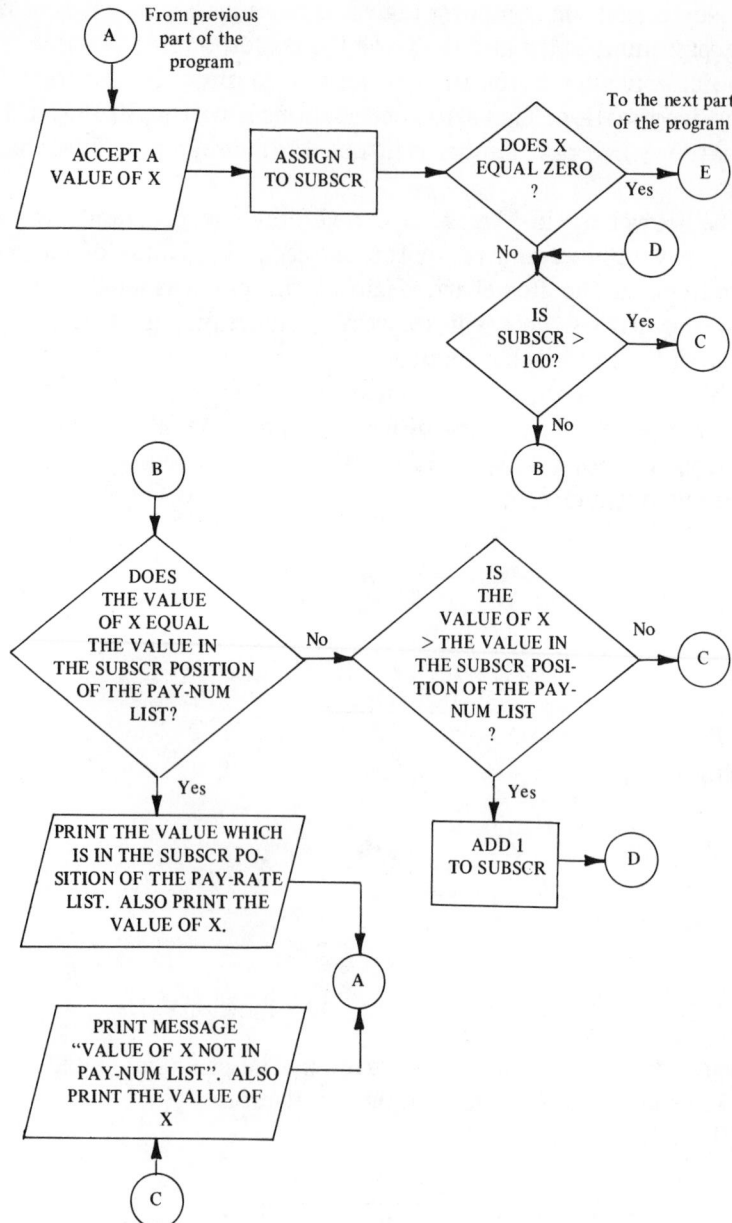

FIG. 11-2

indicates that a calculation is to be made. The calculation is shown within the rectangle.

Finally, the symbol

represents connectors from one part of the program to another.

Now let us examine the flowchart step by step. The first symbol

has an appended legend that tells you that the flowchart is continuing from a point somewhere in the midst of a larger flowchart. In other words, there is a part of the flowchart that is not shown here. Since that part is not needed in order to understand the example given, it is not shown.

Follow the arrow to the symbol

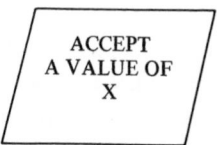

This symbol shows that a program instruction will be given that requires the programmer to type in a number. The computer will accept the number and call it X.

What is X? X represents the pay number for which the program will search in the PAY-NUM list. For purposes of illustration, let us assume that the programmer wishes to know the pay rate corresponding with pay number 12603. When the program indicates its willingness to accept a pay number, the programmer will type 12603.

Follow the arrow to the next symbol:

This portion of the flowchart will call for an instruction that assigns to the name SUBSCR the value 1. This name, which could be any name the programmer wishes to use, will hold a *position* to be examined in the PAY-NUM list. Since we wish to have the computer begin its search by looking at the *first* cell of the PAY-NUM list, SUBSCR is given an *initial* value of 1. Later, SUBSCR's value will change to 2, 3, 4, etc. The largest value that may be held by SUBSCR is 100.

Now follow the arrow to the next symbol:

The two arrows flowing out of this diamond show that there are two possible answers to the question: yes or no. If the answer is yes, the computer will go to the next part of the program (not shown in this flowchart). If the answer is no, the computer will go to the next decision diamond. Exactly why the program checks to see whether X equals zero will be explained later.

The next decision diamond is

The program will ask the question, Is the value of SUBSCR greater than 100? Since SUBSCR will begin with 1, later increase to 2, then 3, etc., it is necessary to have the program check whether its maximum value has been reached. If SUBSCR is larger than 100, the program will jump to Ⓒ , where an error message is printed; if SUBSCR is not greater than 100, the program will go to Ⓑ , where the search for X (pay number 12603) will begin.

When SUBSCR is 1, the computer will go to the next decision diamond:

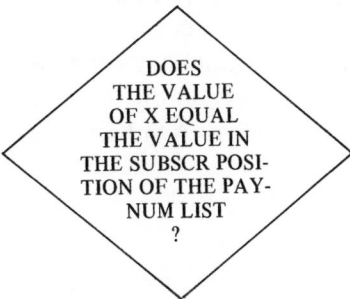

The program will be directed to *compare* the value of X (12603) with the contents of the first location of the PAY-NUM list to see if they are equal. The first value in the PAY-NUM list is 12463; therefore there will be no match. The program will exit from the decision diamond by taking the path labeled No. This path will lead immediately to another decision diamond.

This decision diamond will tell whether or not the search for X (pay number 12603) should continue. Since the pay numbers in the PAY-NUM list are sorted in increasing sequence, X must always be equal to or greater than the current list member being examined. If X were smaller than the current list member being examined, then it would be obvious that the value being sought is not in the list and the search should not continue.

Since 12603 *is* greater than the first member of the PAY-NUM list (12463), the program will take the Yes path.

In the last two decision diamonds, the program concerned itself with the *first position* of the PAY-NUM list. Why? Because SUBSCR had the value 1. If SUBSCR had some other value between 2 and 100, inclusive, the program would have compared X with some other value of the PAY-NUM list. (You have probably noticed that SUBSCR is short for *subscript*. A subscript holds the position of a particular value in a list.)

The next symbol is

```
┌─────────────────────┐
│                     │
│      ADD 1 TO       │
│       SUBSCR        │
│                     │
└─────────────────────┘
```

The value of SUBSCR changes. If it had been 1, it now becomes 2; had it been 50, it now becomes 51; had it been 100, it now becomes 101. The program is then directed to follow the arrow. The arrow leads to ⒟ , which, in turn, directs the program to return to the decision diamond:

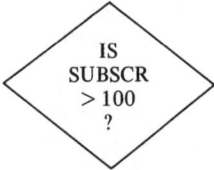

You can see that when SUBSCR is 2, the program will answer the question No, and another examination of the PAY-NUM list will be made. This time, the second value of the PAY-NUM list (12489) will be checked to see if it equals the value assigned to X (12603).

Why is it that the *second* value of the PAY-NUM list will be examined? Because SUBSCR equals 2.

Of course 12489 and 12603 are not equal, so the computer will go to the next decision diamond where a check is made to see whether 12603 is greater than 12489. It is; therefore, the program will change SUBSCR to 3 and continue the search.

Will the program ever find 12603 in the PAY-NUM list? Yes, when SUBSCR's value is 5. Look at the PAY-NUM list. You can see that 12603 is the fifth value of the list. The flowchart shows that when the answer to the decision diamond

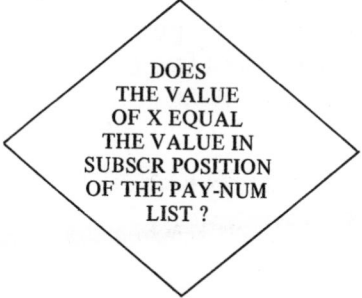

is Yes, the program goes to the symbol

PRINT THE VALUE WHICH IS IN THE SUBSCR POSITION OF THE PAY-RATE LIST. ALSO PRINT THE VALUE OF X.

The computer will be given an instruction that tells it to print the fifth value of the PAY-RATE list ($4.73) and the value of X (12603). Why is it the *fifth* value of the PAY-RATE list that is printed?—because the value 12603 was found in the fifth position of the PAY-NUM list. The computer *remembers* that SUBSCR's value is 5.

What will the computer be told to do *after* it has printed the fifth value in PAY-RATE list? It will be told to return to Ⓐ and *repeat* the task. That is, the computer will await another pay number to be typed into the program and assigned to X. The *new* value of X will replace the *old* value.

Suppose that the new value is 12554. The computer will print the *third* value of the PAY-RATE list ($5.14) and the value of X (12554).

Then the computer will return to Ⓐ , expecting still another value to be assigned to X. Suppose that value is 12618. How far into the PAY-RATE list will the program search until it finds that further searching is futile? The program examines 12463, 12489, 12554, 12593, 12603, 12613, and 12620. In those seven examinations, the program finds that the value of X (12618) does not equal any of the PAY-NUM values looked at. The program quits examining values in the PAY-NUM list because X's value (12618) is not greater than the seventh position of the PAY-NUM list. The program will therefore print

VALUE OF X NOT IN PAY-NUM LIST   12618

What is the value of SUBSCR when the message is printed? The value is 7.

After printing the message, the program goes back to Ⓐ expecting to be given another value of X. Suppose the value given is zero. This is the programmer's way of telling the computer to go to the next part of the program, the part not shown in this flowchart. Recall the symbol that reads

When X *does* equal zero, the computer is directed to go to Ⓔ , which represents the beginning of the next part of the program.

Is the decision diamond

needed? Yes. Suppose that the last value in the PAY-NUM list is 15330 and that the most recently accepted value of X is 15338. The program will examine all 100 cells in the PAY-NUM list without finding 15338. Since the value of X (15338) is found to be larger than the hundreth value of the PAY-NUM list (15330), the program will add 1 to SUBSCR and go back to the above decision diamond. Because SUBSCR will now be greater than 100, the decision diamond will cause the program to stop searching the PAY-NUM list.

Here is a portion of a program written in COBOL that reflects the above flowchart:

```
   :   ⎰ Other parts of the COBOL
   ·   ⎱ program not shown.
```

H1.   ACCEPT X. MOVE 1 TO SUBSCR. IF X
      EQUALS ZERO THEN GO TO H20; OTHERWISE GO
      TO H2.

H2.   IF SUBSCR IS GREATER THAN 100 GO TO H6;
      OTHERWISE GO TO H3.

H3.   IF X EQUALS PAY-NUM(SUBSCR) GO TO H4;
      OTHERWISE GO TO H5.

H4.   DISPLAY PAY-RATE(SUBSCR), X. GO TO H1.

H5.   IF X IS GREATER THAN PAY-NUM(SUBSCR)
      ADD 1 TO SUBSCR THEN GO TO H2; OTHERWISE
      GO TO H6.

H6.   DISPLAY "VALUE OF X NOT IN PAY-NUM
      LIST", X. GO TO H1.

H20.   (This is where the next part of
              the program begins.)

This is not necessarily the best way to write the program in COBOL, but it is an easy way to understand. Compare the program against the flowchart and you will see that they match. The COBOL program tells the computer what to do in accordance with the plan shown in the flowchart.

Perhaps the greatest difficulty in understanding the COBOL program lies in the use of SUBSCR. As we said, SUBSCR tells *which* positions of the named list or lists are to be referenced. In COBOL, subscripts must be enclosed within parentheses following the name of the corresponding list. Thus when SUBSCR has the value 5, PAY-NUM(SUBSCR) refers to the fifth position of PAY-NUM; when SUBSCR has the value 50, it refers to the fiftieth position of PAY-NUM; etc.

Now that we have gone through a flowchart in detail, we can present more flowcharts with the expectation that you will be able to understand them without detailed explanations. Consider the flowchart in Fig. 11-3.

If we could depend on the program's user to type his X values in increasing sequence, the flowchart in Fig. 11-3 could be used. You will notice that, unlike the first flowchart, this flowchart calls for searches to begin at various points of the PAY-NUM list—not always at the beginning. For example, suppose that the programmer wishes to find the pay rates corresponding with pay numbers 12489, 12603, and 12620; the program will initially begin searching at the first location of the PAY-NUM list. It will find 12489 when SUBSCR equals 2. The program will print the required information, then set SUBSCR TO 3, and begin searching for pay number 12603. Thus, instead of beginning the search at the beginning of the PAY-NUM list, the program will begin with the third value. It will find 12603 when SUBSCR equals 5. The program will then set SUBSCR to 6 and begin searching for pay number 12620 beginning with the sixth location of the PAY-NUM list. The program will find 12620 when SUBSCR equals 7. It will then set SUBSCR to 8 and request another input from the typewriter.

Suppose that the programmer makes a mistake by typing a pay number that is not in the PAY-NUM list. The program will discover this fact, print a message accordingly, and then, *without changing* SUBSCR, it will ask the programmer to type in another part number. Presumably, the programmer will discover his error and then type in the correct pay number. The program will then begin searching the PAY-NUM list at exactly the same point where it began its unsuccessful search.

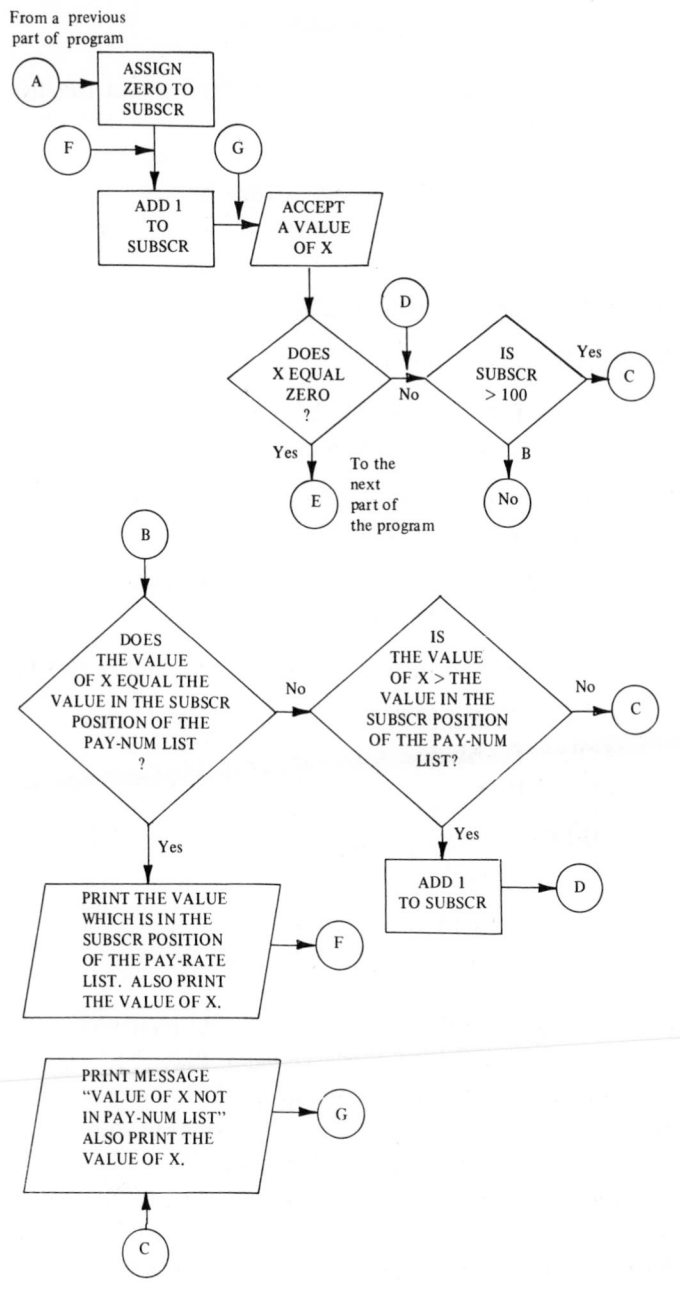

FIG. 11-3

A COBOL program written in accordance with the flowchart in Fig. 11-3 is this one:

$$: \left\{ \begin{array}{l} \text{Other parts of the COBOL} \\ \text{program not shown.} \end{array} \right.$$

H1. MOVE ZERO TO SUBSCR.

H2. ADD 1 TO SUBSCR.

H3. ACCEPT X. IF X EQUALS ZERO GO TO H9;
OTHERWISE GO TO H4.

H4. IF SUBSCR IS GREATER THAN 100 GO TO H8;
OTHERWISE GO TO H5.

H5. IF X EQUALS PAY-NUM(SUBSCR) GO TO H6.
OTHERWISE GO TO H7.

H6. DISPLAY PAY-RATE (SUBSCR), X THEN GO TO H2.

H7. IF X IS GREATER THAN PAY-NUM(SUBSCR)
ADD 1 TO SUBSCR THEN GO TO H4; OTHERWISE
GO TO H8.

H8. DISPLAY "VALUE OF X NOT IN PAY-NUM
LIST", GO TO H3.

H9.   (This is where the next part of
the program begins.)

In the program, DISPLAY is a COBOL instruction that means "print a line on the printer." Observe, also, that at the very beginning of the illustrated routine, the subscript SUBSCR is initialized to zero. This is done so that there is a definite beginning value for SUBSCR. (Not all computers initialize variables to zero.)

The illustrated method of searching a list is fine when we can depend on the program's user to type numbers in increasing sequence. But humans are prone to make mistakes and it would be better to devise a search routine that is efficient, yet does not make rigid demands on the user.

A method that always begins searching at the beginning of a list when a random series of numbers is given to the program is too inefficient. On the average, half of the list will have to be examined to find any given number. Yet we may want to enable the user to enter his numbers in random order. What can be done?

The "binary" search method is the answer. This method works for any size list of numbers; in fact, the larger the list, the more efficient it is. For illustration, consider a list containing 19 catalog numbers. While this is a rather short list, the binary search method uses the same principles as it does for much longer lists. The 19-member catalog number list is given in Fig. 11-4. The positions of the 19 catalog numbers have been given for reference. They are not actually stored within the 19 cells. Note that the catalog numbers stored in the list are in increasing sequence. They *must* be.

CAT	POSITION
346	1
349	2
356	3
363	4
378	5
383	6
385	7
389	8
395	9
406	10
421	11
423	12
443	13
453	14
475	15
478	16
486	17
498	18
504	19

FIG. 11-4

Suppose that it is desired to find a catalog number known to be in this CAT list. The program will look first in the middle cell of the list (cell 10). Since the catalog numbers are in increasing sequence, the catalog number actually stored in cell 10 tells whether the catalog number to be found is in cell positions 1 through 9 or 11 through 19. (Of course, if catalog 406 were required, it would be found at once in cell position 10 of the list.)

Once the program knows whether the catalog number to be found is in the first half of the CAT list or the second half, it divides that half in two and examines the contents of the middle cell there.

Let us try an actual example. Suppose that the value to be found is catalog 385. The computer looks first at cell 10 of the list. The catalog number in cell 10 is 406. Since 406 is larger than 385, the program knows that the number to be found must be located somewhere within a cell at positions 1 through 9. The program looks next at position 5 of the CAT list. The catalog number there is 378. Since 378 is smaller than 385, the program deduces that the value to be found, 385, must lie within a cell at positions 6 through 9. The computer will, therefore, examine next either cell 7 or 8. Suppose that it examines cell 8. The value sought, 385, is not there, but since the value at position 8, 389, is larger than 385, the computer deduces that the value to be found must be in cell 6 or 7.

The program converges rapidly to the cell where the sought-for catalog numbers must be. Later, we shall inspect a table that shows how many cell examinations are necessary to find numbers in lists of various sizes.

It is clear that a program employing the binary search method must be carefully planned. The method must be able to find any number in a list in the quickest way possible and it must be able to determine when a sought-for number is not in the list. Study the binary search flowchart given in Fig. 11-5.

A COBOL program which agrees with the flowchart in Fig. 11-5 is

:   { Instructions not shown. From a previous
:   { part of the program.

D1.   ACCEPT X. If X = ZERO GO TO D5;
       OTHERWISE MOVE 1 TO L MOVE 19 TO H
       THEN GO TO D2.

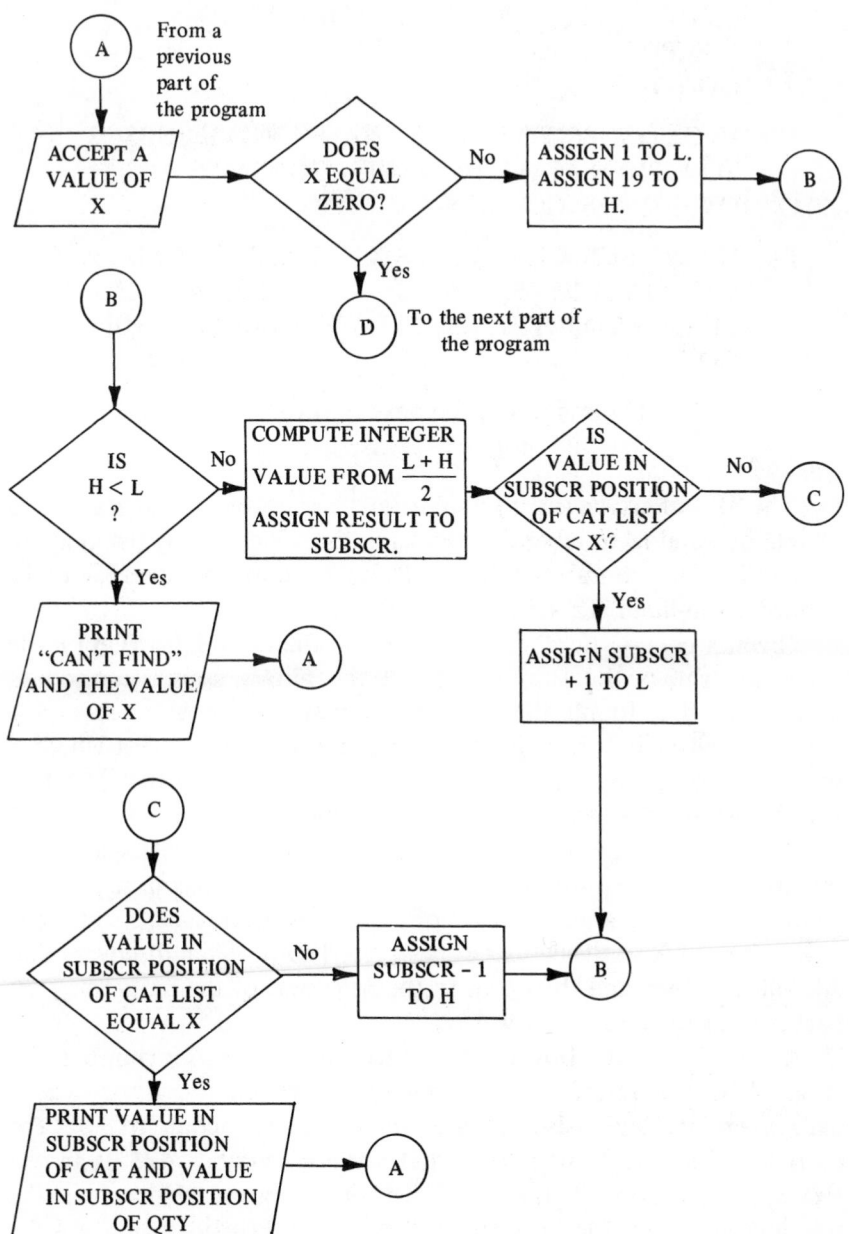

FIG. 11-5

D2. IF H IS LESS THAN L DISPLAY "CAN'T
FIND", X THEN GO TO D1; OTHERWISE
COMPUTE SUBSCR = (L + H)/2.
GO TO D3.

D3. IF CAT(SUBSCR) IS LESS THAN X MOVE SUBSCR
TO L THEN ADD 1 TO L THEN GO TO D2;
OTHERWISE GO TO D4.

D4. IF CAT(SUBSCR) EQUALS X DISPLAY CAT(SUBSCR),
QTY(SUBSCR) THEN GO TO D1; OTHERWISE MOVE
SUBSCR TO H THEN SUBTRACT 1 FROM H THEN
GO TO D2.

D5.   (This is where the next part of
the program begins.)

The flowchart covers the solution to a problem that involves a
search of catalog numbers in a catalog list. The catalog list that we
are calling CAT contains 19 items. There is a corresponding list of 19
quantities-on-hand. This list is called QTY.

Given a catalog number, X, to find in the catalog list (CAT), the
flowchart shows the employment of the binary search method to
find it. Having found the catalog number, the program saves the
value of SUBSCR, which gives the *position* of the CAT list where X
was found. This value is used to obtain the quantity (in the QTY list)
which corresponds to the catalog item found.

After the program has found a value of X and given a report
about it, the computer is told to return to the beginning of the
routine to obtain another value of X. The program keeps obtaining
new values of X until the user types zero. The program discovers that
X's value is zero and then goes to the next part of the program. That
part is not shown in the flowchart.

Let us follow the flowchart as it tries to find catalog number 478
(Fig. 11-6). For reference, we repeat here the contents of the catalog
list given earlier. Also shown now is the contents of the
corresponding QTY list. Note that catalog number 478 is in the
sixteenth position of the CAT list. This means that when the
problem is solved, the value of SUBSCR is 16 and the values of CAT
and QTY printed are 478 and 55, respectively.

POSITION	CAT	QTY
1	346	18
2	349	3
3	356	0
4	363	44
5	378	91
6	383	17
7	385	13
8	389	2
9	395	39
10	406	48
11	421	14
12	423	86
13	443	51
14	453	7
15	475	26
16	478	55
17	486	37
18	498	70
19	504	33

FIG. 11-6

The value assigned to X is 478.

478

X

The values assigned to L and H (L means "low" and H means "high") are 1 and 19, respectively:

H is not less than L; therefore SUBSCR is computed from the integer portion of (L + H)/2. This is (1 + 19)/2 or 10. (Note that the integer portion of a calculation is the whole-number portion. Fractions, if any, are dropped.)

The value in the tenth position of the CAT list (406) *is* less than X (478); therefore SUBSCR + 1 (11) is assigned to L. The new value of L *replaces* the old one. The values of L and H are now

At this point the computer knows that the value of X being sought (478) must be within CAT list locations 11 through 19, inclusive. Therefore L is set to 11 and H remains at 19. (L gives the lowest cell location where X could be found and H gives the highest.)

The program goes back to Ⓑ , where H and L are tested against each other. H is not less than L; therefore SUBSCR is computed again. This time SUBSCR is the integer portion of (11 + 19)/2 or 15.

┌──────┐
│ 15 │
└──────┘
SUBSCR

The new value of SUBSCR replaces the old one.

The value at the fifteenth position of CAT is 475. This value *is* less than the value being sought (478); therefore SUBSCR + 1 (16) is assigned to L. The new value of L *replaces* the old one. The values of L and H are now

16   19

L   H

Again the program goes back to Ⓑ . H is not less than L; therefore SUBSCR is computed again. This time it is the integer portion of (16 + 19)/2 or 17.

Now SUBSCR is 17:

17

SUBSCR

The new value of SUBSCR replaces the old one.

This time the value in the seventeenth position of the CAT list (486) is greater than the value being sought (478); therefore the program flow goes to Ⓒ , where the computer is directed to check whether 486 and 478 are equal. They are not, so SUBSCR-1 (16) is assigned to H. The new value of H replaces the old one. The values of L and H at this point are

16   16

H   L

The program has zeroed in on the exact location of catalog 478. The next time SUBSCR is computed, it is 16. The program is asked whether the value located at the sixteenth position of the CAT list is less than the value of X. It is not (both values are 478). The program asks whether the value at the sixteenth position of the CAT list is equal to the value of X. It is, and the program prints the values found at the sixteenth positions of the CAT and QTY lists.

The program then goes back to (A) to obtain another value of X. For practice you should follow the flowchart to see what happens when the value assigned to X is not in the CAT list. Try, for example, catalog number 365.

We suggest you set up a work sheet with boxes like this:

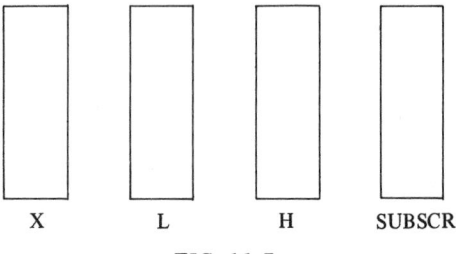

FIG. 11-7

Then as you assign initial or replacement values for the variables, you jot them down in the boxes and cross off the old values. When you are done with this problem, your work sheet should look like this:

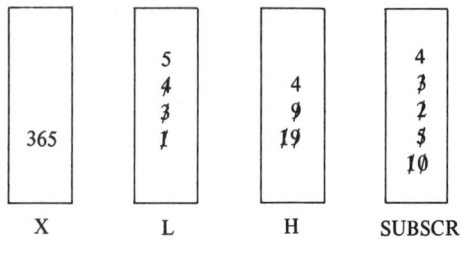

FIG. 11-8

Here is a table showing how many compares are necessary to find a given value in various- size lists:

List Size	Maximum Number of Compares
10	4
100	7
1,000	10
10,000	14

As you can see, the larger the list, the more impressive the efficiency of the method.

You can determine the maximum number of compares necessary to find a value by noting the size, S, of the list. Use this table:

List Size	Maximum Number of Compares
$2^0 \leqslant S < 2^1$	$0 + 1 = 1$
$2^1 \leqslant S < 2^2$	$1 + 1 = 2$
$2^2 \leqslant S < 2^3$	$2 + 1 = 3$
$\vdots$	
$2^{n-1} \leqslant S < 2^n$	$(n - 1) + 1 = n$

Suppose that the size of the list is 5000, for example. The value 5000 lies between 2^{12} (4096) and 2^{13} (8192). The number of compares needed are 13. If the size of the list had been exactly 8192, the table would show that 14 compares would be needed. The line of the table giving this information reads

List Size	Maximum Number of Compares
$2^{13} \leqslant S < 2^{14}$	$13 + 1 = 14$

2^{13} is exactly 8192.

EXERCISES

11-1. When using lists why is it often desired to know the position that a certain value has in a list?

11-2. What is a flowchart?

11-3. In flowcharting what is the function of the diamond-shaped symbol?

11-4. What are the fewest numbers of arrows coming from a diamond-shaped symbol? Why?

11-5. In flowcharting, what are the shapes of the symbols representing the reading of input data, the processing of data, and the printing of output information?

11-6. What is the role of the subscript when a list is searched?

11-7. In a COBOL program, how is the name of a list subscripted? That is, where is the subscript located in relation to the list name?

11-8. In a COBOL program, what is the beginning word of the instruction that causes a decision to be made.

11-9. What are the two ways that a computer program may deduce the fact that a value being sought in a list is not in the list?

11-10. Discuss how the binary search method works.

11-11. Using the binary search method, how many compares are needed, maximum, to find a given value in a list containing 253,000 values?

11-12. Construct a list of 25 values sorted in increasing sequence. Then assign to X the value located at the seventh position of the list. Show how the values of L, H, and SUBSCR change as a program, using the binary search method, finds the value.

11-13. Using the same list in in Exercise 11-12, assign a value to X, one that is *not* in the list. Show how the values of L, H, and SUBSCR change as a program, using the binary search method, deduces that the value is not in the list.

11-14. Using the binary search method, what is the maximum number of compares that a program may make when searching for a value in a list of 25 values?

11-15. Why is it important that the values in a list be in increasing or decreasing sequence in order that the binary search method may work properly?

Chapter 12

SORTING

In processing data, it is often necessary to place a series of numbers, words, names, catalog numbers, etc., in some sequence. For instance, suppose that we have six numbers, 28.3, 1.6, -3.2, 0, 7.6, and -8.1, and that it is desired to place them in increasing sequence. That sequence would be -8.1, -3.2, 0, 1.6, 7.6, 28.3. If it were desired to place those numbers in descending sequence, that sequence would be 28.3, 7.6, 1.6, 0, -3.2, -8.1.

Of course, in an actual data processing application, it is unlikely that you would be dealing with only five numbers. The chances are you would be dealing with several hundred it not thousands of numbers.

Consider another example. Suppose that you had to place the following six names in sequence: WILLIAMS, JOHNSON, ADAMS, WILTS, JOHNS, and JONES. Where names are concerned, increasing sequence means alphabetical order. Therefore the above names would be ordered like this: ADAMS, JOHNS, JOHNSON, JONES, WILLIAMS, WILTS. As you can see, JOHNS is "smaller" than JOHNSON because the first five letters are common to both names, but when the letters in the first name run out, there still are some letters left in the second name. WILLIAMS is "smaller" than WILTS. The first three letters in both names are identical, but the fourth letter in each name determines which name is smaller. Since the second L in Williams is smaller (alphabetically) than the T in WILTS, the name WILLIAMS is considered smaller than the name WILTS.

Sometimes data items contain a mixture of letters and numeric digits. Most computer systems consider numeric digits "smaller" than letters. Consider these catalog numbers: 3X5583, P186Q, 2365D, 21453A, 463L9D, and 3XD55. Increasing sequence would be considered to be this: 21453A, 2365D, 3X5583, 3XD55, 463L9D, P186Q.

It can be seen that 3X5583 and 3XD55 both begin with 3X. Where only 3X is considered, no decision can be made regarding whether 3X5583 or 3XD55 is larger. However, the third character in each catalog number decides the issue.

Following the same reasoning, it can be determined that 463L9D is smaller than P186Q because the first character in 463L9D (4) is smaller than the first character in P186Q (P).

While it is possible to place into sequence (sort) values that contain digits, letters, and other characters, we shall consider only numeric values in this chapter and in Chapter 13.

In Chapter 11 you saw that in order to efficiently perform certain tasks, various lists of numbers had to be in sequence. Searching a list of numbers, for example, could be done very efficiently using the binary search method if the list were arranged in increasing sequence. If the list had been organized in some random fashion, the binary search method would not work. Often, therefore, it is necessary to sort a list of numbers before some task may be successfully performed. Consider the list in Fig. 12-1. Assume that

Positions	ACCT-NUMS
1	441
2	475
3	319
4	484
5	314
6	453
7	436
8	425
9	473
10	467
11	444

FIG. 12-1

this list represents 11 account numbers identifying 11 bank customers. Let us state again that while the example deals only with a small series of numbers, in an actual data processing application you might have thousands. But, what works for 11 numbers should work for thousands.

The account numbers in the ACCT-NUMS list are, at present, in no special sequence, but suppose that it is desired to place them in increasing sequence. One way to have the computer sort the numbers is the *interchange method*. In this method, the list values are examined in pairs. Where the numbers are already in sequence, they are left alone; where they are out of sequence, the values are interchanged.

Consider the pair of numbers at positions 1 and 2 of the ACCT-NUMS list. Since 441 is less than 475, they are already in increasing sequence and are, therefore, left alone.

Now consider the pair of account numbers at positions 2 and 3 of the list. Those account numbers are 475 and 319. Since 475 is not smaller than 319, the two numbers must be interchanged. Value 319 will be moved to position 2 of the ACCT-NUMS list and value 475 will be moved to position 3 of the list.

After the "swap" has been made the ACCT-NUMS list will look as in Fig. 12-2. The next pair is formed of the account numbers at positions 3 and 4 of the list. They *are* in sequence, so no swap is

Positions	ACCT-NUMS
1	441
2	319
3	475
4	484
5	314
6	453
7	436
8	425
9	473
10	467
11	444

FIG. 12-2

made. However, when the pair of numbers at positions 4 and 5 is considered, the necessity for a swap is indicated. The value 314 will be moved to position 4 and the value 484 will be moved to position 5.

The interchange method calls for the pairs of numbers in Fig. 12-3 to be examined. You can see that in a list of 11 numbers, ten pairs of numbers are to be compared. Most of the numbers of a list are members of two different pairs. For instance, the account number at position 2 of the ACCT-NUMS list is examined when the first pair is processed and also when the second pair is processed.

First Position of Pair	*Second Position of Pair*
1	2
2	3
3	4
4	5
5	6
6	7
7	8
8	9
9	10
10	11

FIG. 12-3

After the ten pairs of numbers have been examined and all necessary interchanges have been made, the ACCT-NUMS list will look as in Fig. 12-4. You can see that the account numbers have not

Positions	ACCT-NUMS
1	441
2	319
3	475
4	314
5	453
6	436
7	425
8	473
9	467
10	444
11	484

FIG. 12-4

yet been sorted. What, then, has been accomplished by the processing of the ten pairs of numbers? Well, we can say that the *largest* number of the list has been forced to fall to the bottom of the list. But it is obvious that at least another pass must be made through the ACCT-NUMS list. This time, though, we need to examine only the first nine pairs of numbers. The reason is that we know for certain that the largest number has fallen to the bottom of the list. There is no necessity, therefore, to look at this number again.

During the second pass, therefore, the computer will have to examine only the pairs in Fig. 12-5. During the second pass, nine pairs of numbers will be examined and swaps made where necessary.

First Position of Pair	Second Position of Pair
1	2
2	3
3	4
4	5
5	6
6	7
7	8
8	9
9	10

FIG. 12-5

After the pass has been completed, the ACCT-NUMS list will look as shown in Fig. 12-6. You can see that the second largest account number has fallen to a position near the bottom of the list, to the

Positions	ACCT-NUMS
1	319
2	441
3	314
4	453
5	436
6	425
7	473
8	467
9	444
10	475
11	484

FIG. 12-6

second from the last position. But the account numbers are not yet sorted. More passes through the ACCT-NUMS list will be required. To be exact, up to eight more passes may be required. Each pass will call for the examination of fewer and fewer pairs. Figure 12-7 shows the number of pairs required for each of the remaining passes required to sort these numbers.

Pass	Pairs to Examine
3	8
4	7
5	6
6	5
7	4
8	3
9	2
10	1

FIG. 12-7

Let us see what happens during the third pass. Figure 12-8 shows that eight pairs of account numbers will be examined. After the third pass has been completed, the ACCT-NUMS list will look as shown in Fig. 12-9.

First Position of Pair	Second Position of Pair
1	2
2	3
3	4
4	5
5	6
6	7
7	8
8	9

FIG. 12-8

Positions	ACCT-NUMS
1	319
2	314
3	441
4	436
5	425
6	453
7	467
8	444
9	473
10	475
11	484

FIG. 12-9

The list is beginning to have a sorted look but obviously more passes will be necessary. It turns out that only two more passes are needed to sort the ACCT-NUMS list completely. Any program that employs this method of sorting a list should include a way of detecting that the list has been sorted despite the fact that the maximum number of passes has not been executed.

Figure 12-10 is a summary of the original order of the 11 account numbers of this example and how that order was improved by each of five passes. Passes 6 through 10 were not necessary in this example because the numbers became sorted earlier than they would have if the original order had been worse. Under the worst situation, ten passes would have been necessary, one pass less than the number of account numbers in the list.

Original Order	Pass 1	Pass 2	Pass 3	Pass 4	Pass 5
441	441	319	319	314	314
475	319	441	314	319	319
319	475	314	441	436	425
484	314	453	436	425	436
314	453	436	425	441	441
453	436	425	453	453	444
436	425	473	467	444	453
425	473	467	444	467	467
473	467	444	473	473	473
467	444	475	475	475	475
444	484	484	484	484	484

FIG. 12-10

The flowchart in Fig. 12-11 shows how the interchange method of sorting could be programmed. In this flowchart, N counts down the maximum number of passes actually required to solve the problem. Since there are 11 numbers in the list, its original setting is 10. You will notice that at the end of each pass, 1 is subtracted from N. When N equals zero, ten passes have been performed and the computer will go to the next part of the program.

J is the name of the subscript that helps form the pairs of numbers to be examined. A pair of account numbers to examine is formed by appending the subscript J and J + 1 to the name of the list.

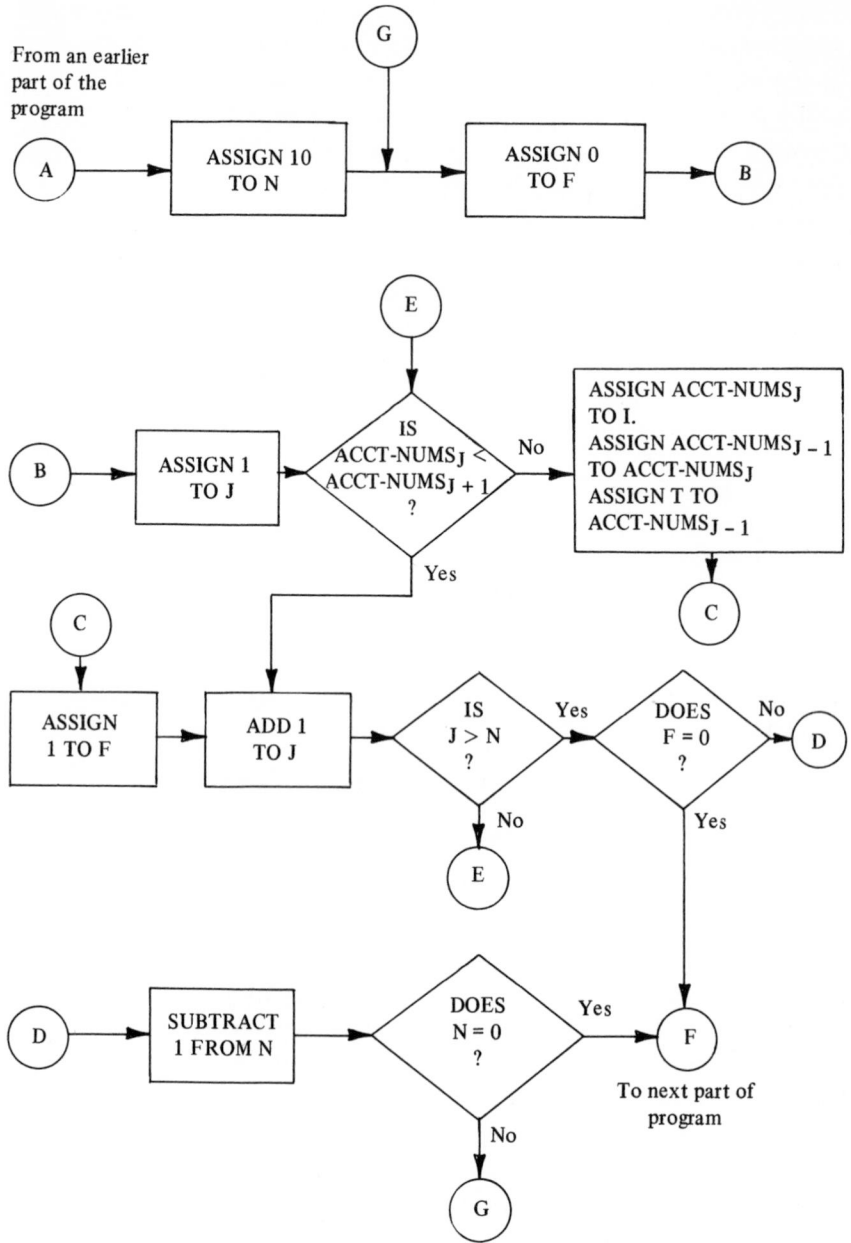

FIG. 12-11

Observe that the value of J increases by 1 after each pair examination. The flowchart shows that J will never be greater than N. During the first pass, therefore, J's largest value will be 10. Since J and J + 1 help form pairs to examine, the last pair of account numbers looked at will be the tenth and eleventh numbers. (At that time J will be 10 and J + 1 will be 11.)

The variable F is a *telltale flag*. It tells whether or not the last pass caused *any* interchange to take place. If even one interchange took place during the pass, another pass will be required. F's initial setting is zero. If a pass is completed and F's value did not change, the pass was completed without an interchange taking place. The program, therefore, goes to the beginning point of the next part of the program. If F's value did change from zero to 1, an interchange did take place and another pass is needed (unless, of course, that interchange occurred during the last scheduled pass of the program—in the example, during the tenth pass).

Study where F's value is initially assigned the value zero, where F's value is changed to 1, and where F's value is tested. Observe, also, that F's value is always reset to zero at the beginning of any pass.

Whenever a pair of account numbers is examined, the program decides whether to interchange those values or to increase J by 1 and examine the next pair of numbers. If the program decides to interchange the numbers, a temporary working cell, T, is needed to prevent the loss of one of the numbers in the pair. Suppose, for example, that it is required to interchange the values at list positions 5 and 6 (Fig. 12-12). It would obviously be wrong to place the

Positions	ACCT-NUMS
5	436
6	425

FIG. 12-12

contents of cell 6 into cell 5 and then place the contents of cell 5 into cell 6. *Both cells* would then hold account number 425. A workable plan is to place the contents of cell 6 into a cell labeled T, then place the contents of cell 5 into cell 6, and, finally, place the contents of cell T into cell 6 (Fig. 12-13). This procedure will accomplish the required result: to interchange the contents of cells 5 and 6. Study steps 1, 2, and 3 in Fig. 12-13. Is this the only way to accomplish the task?

The cell labeled T is a temporary holding cell and is used whenever *any* interchange is to be made involving *any* pair of the list during *any* pass.

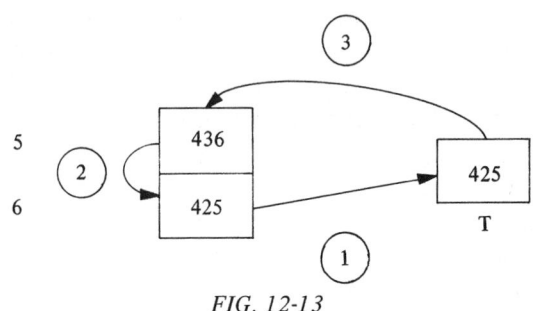

FIG. 12-13

A COBOL program that agrees with the flowchart in Fig. 12-11 is this:

> : ⎧ Statements from an earlier
> · ⎩ part of the program not shown.

P1. MOVE 10 TO N.

P2. MOVE ZERO TO F. MOVE 1 TO J.

P3. IF ACCT-NUMS(J) IS LESS THAN ACCT-NUMS(J+1) GO TO P5; OTHERWISE MOVE ACCT-NUMS(J) TO T THEN MOVE ACCT-NUMS(J+1) TO ACCT-NUMS(J) THEN MOVE T TO ACCT-NUMS(J+1).

P4. MOVE 1 TO F.

P5. ADD 1 TO J. IF J IS GREATER THAN N GO TO P6; OTHERWISE GO TO P3.

P6. IF F EQUALS ZERO GO TO P8; OTHERWISE SUBTRACT 1 FROM N.

P7. IF N EQUALS ZERO GO TO P8; OTHERWISE GO TO P2.

P8.   (Program continues from this point.)

A better way to sort values in a list is the *Shell method,* developed by Dr. Donald L. Shell of the Corporate Research & Development Center, General Electric Co., Schenectady, N.Y.

To begin with, let us consider the 11 original unsorted values of the ACCT-NUMS list (Fig. 12-14). These values will be sorted in much the same way that the interchange method uses. There are two important differences:

1. There are fewer passes through the list.

2. Each pass through the list deals with pairs of numbers that are not necessarily in consecutive cells.

Positions	ACCT-NUMS
1	441
2	475
3	319
4	484
5	314
6	453
7	436
8	425
9	473
10	467
11	444

FIG. 12-14

The number of passes through the list is determined by its size. Use the table shown in Fig. 12-15. Since the ACCT-NUMS list contains 11 values, the table shows that the row labeled $2^3 \leqslant S < 2^4$ applies. That row shows that three passes are necessary to sort the list.

Each pass through the list deals with cells that may or may not be in consecutive locations. During the earlier passes, the cells of the numbers being examined are either exactly or almost half the list distant from each other. As the passes are performed, the distance between the numbers becomes smaller and smaller, During the last pass, the numbers being examined are in consecutive cells.

Size of List	Maximum Number of Passes Necessary
$2^0 \leqslant S < 2^1$	0
$2^1 \leqslant S < 2^2$	1
$2^2 \leqslant S < 2^3$	2
$2^3 \leqslant S < 2^4$	3
.	.
.	.
.	.
.	.
$2^{n-1} \leqslant S < 2^n$	n-1

FIG. 12-15

In this example, the program will first compare the account number in the first location of the list with the account number in the sixth location. You can see that cell 6 is at the exact midpoint of the list. The values being examined are 5 cells apart, approximately half the size of the list.

If the values being examined are in sequence, the program will leave them alone; if not, the program will interchange them. In the example, the two account numbers are 441 and 453. They are in sequence; therefore no action is taken.

Now the program will examine the second and seventh values. (The cells are still five apart.) These are not in sequence; therefore the program will interchange them. After the interchange has taken place the ACCT-NUMS list will look as shown in Fig. 12-16.

Positions	ACCT-NUMS
1	441
2	436
3	319
4	484
5	314
6	453
7	475
8	425
9	473
10	467
11	444

FIG. 12-16

In like manner, the program will examine the values at positions 3 and 8, 4 and 9, 5 and 10, and 6 and 11. The distance between account numbers throughout these six examinations will remain constant at five cells. Whenever the two values being examined are in sequence, no action will be taken; whenever the values are not in sequence, they will be interchanged.

The program has made the *first* pass through the ACCT-NUMS list. Two more passes are needed. (If the size of the list were greater than 11, *more* than two additional passes might be necessary, as you have already seen.)

At the end of the first pass, the ACCT-NUMS list will look as shown in Fig. 12-17. A number of interchanges will have been made. Note especially that the account number at position 6 has been examined twice.

Positions	ACCT-NUMS
1	441
2	436
3	319
4	473
5	314
6	444
7	475
8	425
9	484
10	467
11	453

FIG. 12-17

Let us determine the rules that govern what values will be examined during the first pass. The size of the list is divided by 2. If a fraction results, it is disregarded. Call the result I (for interval). When 11 is divided by 2, the integer answer is 5. This is I. I gives the interval between values being examined. During the first pass, the values to be examined are at positions

1 and 6

2 and 7

3   and   8

4   and   9

5   and   10

6   and   11

To prepare for the second pass, another I must be computed. This is done by dividing the ACCT-NUMS list size by 4. The integer result (2) becomes I. (As before, fractions are disregarded.) The positions of the values to be examined are

1   and   3

2   and   4

3   and   5

4   and   6

5   and   7

6   and   8

7   and   9

8   and   10

9   and   11

To prepare for the first pass, the divisor to apply to the list size was 2. To prepare for the second pass, the divisor was twice 2 or 4. To prepare for the third pass, the divisor to apply to the list size will be twice 4 or 8. Where additional passes are required, new divisors are obtained by doubling the value of the *last* divisor used.

Now let us go through the second pass in detail. The first two account numbers to examine are the ones located at list positions 1 and 3. These values are 441 and 319. Since they are not in sequence, they are interchanged.

The values at positions 2 and 4 are examined next. Those values are 436 and 473. Since they are in sequence, no action is taken. Next, the values at positions 3 and 5 are examined. The values are 441 and 314. Being out of sequence, they are interchanged (bear in mind that position 3 contains account number 441, not 319, because of an earlier interchange).

At this point the ACCT-NUMS list looks as shown in Fig. 12-18. Now, instead of going forward to examine the account numbers at positions 4 and 6, the program *backs up* to reexamine positions 1 and 3. The swap of values at positions 3 and 5 *triggered* the recheck of the previously examined values.

Positions	ACCT-NUMS
1	319
2	436
3	314
4	473
5	441
6	444
7	475
8	425
9	484
10	467
11	453

FIG. 12-18

Note that the values at positions 1 and 3 are out of sequence and must, therefore, be interchanged. If they had been in sequence, no action would have been required.

In general, we may observe that when an interchange takes place the program *backs up* to recheck previously examined values. Backing up takes place as far back as possible, to the very beginning of the list if need be, so long as values keep getting interchanged. Whenever two values are examined and an interchange *does not* take place, backing up ceases, and the program advances to examine the next two values in the list.

Following this plan, the second pass through list ACCT-NUMS is completed. At that time the list will look as shown in Fig. 12-19. The program is now ready to go into the third pass. First the interval between list positions, I, must be computed. The divisor is 8. The integer value that results when 11 is divided by 8 is 1. Therefore, during the final pass through the ACCT-NUMS list, the pairs of values to be examined will be in consecutive sequence. These are the pairs the program will examine:

$$1 \text{ and } 2$$

$$2 \text{ and } 3$$

$$3 \text{ and } 4$$

$$4 \text{ and } 5$$

$$5 \text{ and } 6$$

$$6 \text{ and } 7$$

$$7 \text{ and } 8$$

$$8 \text{ and } 9$$

$$9 \text{ and } 10$$

$$10 \text{ and } 11$$

Positions	ACCT-NUMS
1	314
2	425
3	319
4	436
5	441
6	444
7	453
8	467
9	475
10	473
11	484

FIG. 12-19

During this pass, where backing up is necessary, the interval between list positions will be 1. In fact, whenever any backing up is done, the interval between cells is the same as the interval used when moving forward.

If you study the action that takes place during the third pass, you will see that the first and second passes did such a good job of ordering the account numbers that only two interchanges take place during the pass and *no backing up at all* is needed. The final order of the values in the ACCT-NUMS list is as shown in Fig. 12-20.

Positions	ACCT-NUMS
1	314
2	319
3	425
4	436
5	441
6	444
7	453
8	467
9	473
10	475
11	484

FIG. 12-20

A flowchart that defines the Shell sorting method is as shown in Fig. 12-21. In this flowchart the meanings of the symbols used are as follows:

D   Divisor. D's initial value is 1 but it is doubled at once. Therefore D's value, when it is first used to divide the size of the list, is 2.

I   Interval size. I's value is initially computed as 5, then 2, then 1, and finally zero. The program knows that the list has been sorted when I's value becomes zero.

L   Lower of the two list-position subscripts, i.e., the smaller of the two position subscripts.

H   Higher of the two list-position subscripts, i.e., the larger of the two position subscripts.

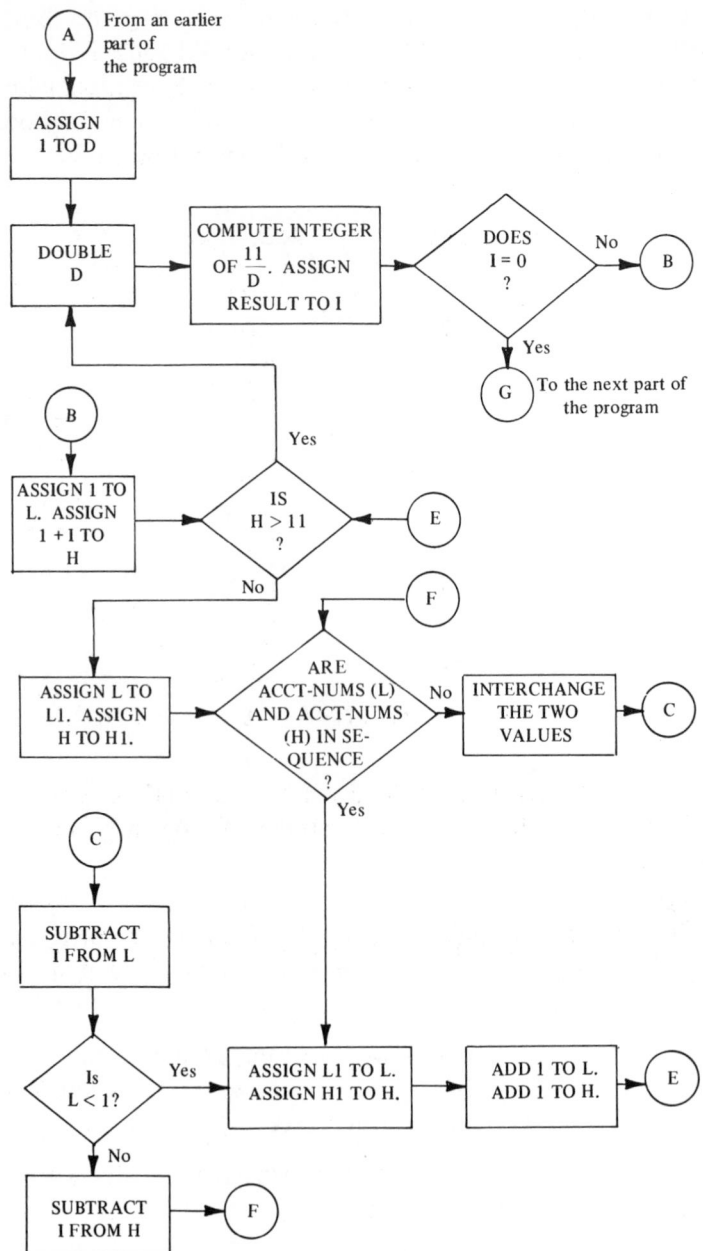

FIG. 12-21

L1   A temporary holding cell for L. L's value is saved so that if backing up takes place, the program will later be able to resume its forward motion at the correct place.

H1   A temporary holding cell for H. H's value is saved for the same reason that L's value is saved.

Here is a COBOL program that agrees with the flowchart in Fig. 12-21 (being careful not to confuse I with 1):

$\left.\begin{array}{l} \vdots \\ \vdots \end{array}\right\{$ Statements from an earlier part of the program not shown.

T1.   MOVE 1 TO D.

T2.   MULTIPLY 2 BY D. COMPUTE I EQUALS 11/D.
IF I EQUALS ZERO GO TO T11; OTHERWISE
GO TO T3.

T3.   MOVE 1 TO L. MOVE I TO H. ADD 1 TO H.

T4.   IF H IS GREATER THAN 11 GO TO T2;
OTHERWISE GO TO T5.

T5.   MOVE L TO L1. MOVE H TO H1.

T6.   IF ACCT-NUMS(L) IS LESS THAN ACCT-NUMS(H)
GO TO T10; OTHERWISE GO TO T7.

T7.   MOVE ACCT-NUMS (L) TO T.
MOVE ACCT-NUMS (H) TO ACCT-NUMS(L).
MOVE T TO ACCT-NUMS(H).

T8.   SUBTRACT I FROM L. IF L IS LESS THAN 1 GO
TO T10; OTHERWISE GO TO T9.

T9.   SUBTRACT I FROM H. GO TO T6.

T10.   MOVE L1 TO L. MOVE H1 TO H.
ADD 1 TO L. ADD 1 TO H. GO TO T4.

T11.   (Program continues from this point.)

It would be educational for the student to "hand-check" the flowchart. A good place to begin is with pass 2. The student should

Positions	ACCT-NUMS
1	441
2	436
3	319
4	473
5	314
6	444
7	475
8	425
9	484
10	467
11	453

FIG. 12-22

copy the list ACCT-NUMS as it appeared at the end of pass 1 (Fig. 12-22). Then draw six rectangles and label them D, I, L, H, L1, and H1 as in Fig. 12-23. The white spaces in Figs. 12-22 and 12-23 are working spaces. You will be writing numbers within those spaces, crossing them off, and replacing them with new numbers.

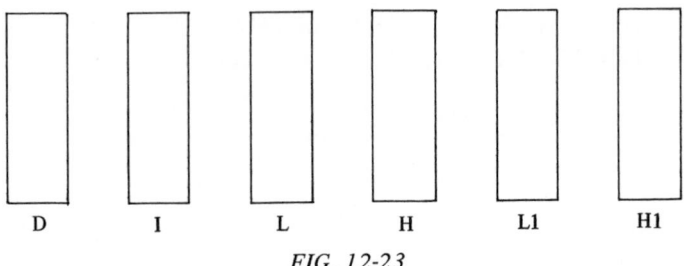

D   I   L   H   L1   H1

FIG. 12-23

Next, begin on the flowchart with the symbol labeled

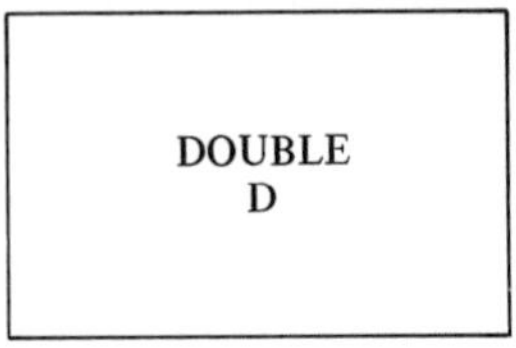

DOUBLE
D

At the end of pass 1, the value of D was 2. D now becomes 4. Write this value in the box labeled D. Proceed to the next symbol in the flowchart. This calls for a calculation of I. The integer value of I is 2. Write this number in the rectangle labeled I. I does not equal zero; therefore the program continues to the flowchart symbol calling for the assignment of 1 to L and 1 + I to H. Write 1 in the rectangle labeled L and 3 in the rectangle H. Since H is not greater than 11, the program proceeds to the flowchart symbol that calls for the assignment of L to L1 and H to H1. Copy the values 1 and 3 into rectangles L1 and H1. Your six rectangles should now look as shown in Fig. 12-24.

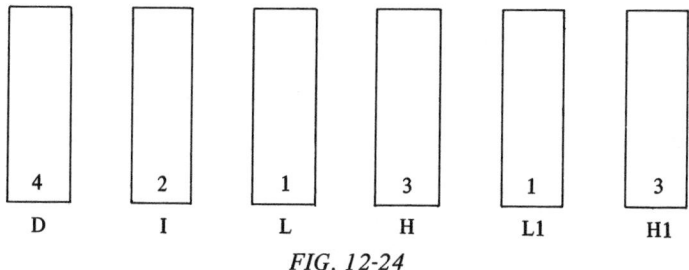

4	2	1	3	1	3
D	I	L	H	L1	H1

FIG. 12-24

Now continue with the flowchart. Whenever the program calls for a change to be made either in the ACCT-NUMS list or in the six rectangles labeled D, I, L, H, L1, and H1, make those changes in the working spaces allowed. For example, after the account numbers at positions 1 and 3 have been interchanged, your working spaces showing the ACCT-NUMS list should look as shown in Fig. 12-25.

Positions	ACCT-NUMS	
1	~~441~~	319
2	436	
3	~~319~~	441
4	473	
5	314	
6	444	
7	475	
8	425	
9	484	
10	467	
11	453	

FIG. 12-25

Positions	ACCT-NUMS			
1	~~441~~	~~319~~	314	
2	~~436~~	425		
3	~~319~~	~~441~~	~~314~~	319
4	~~473~~	~~444~~	~~425~~	436
5	~~314~~	441		
6	~~444~~	~~473~~	~~425~~	444
7	~~475~~	453		
8	~~425~~	~~473~~	467	
9	~~484~~	~~453~~	475	
10	~~467~~	473		
11	~~453~~	484		

FIG. 12-26

Proceed carefully through the flowchart making changes in the working areas as required (Fig. 12-26). When the second pass has been completed (when H *is* greater than 11), your working spaces should contain the notations shown in Fig. 12-27. The rightmost values in the ACCT-NUMS working space give the final arrangement of those values after pass 2 has been completed.

That final arrangement is as shown in Fig. 12-28. The uppermost values in the working spaces labeled D, I, L, H, L1, and H1 give the final values found to those names after pass 2 has been completed. If you continue using these working spaces into pass 3, the flowchart will cause changes beginning with D. D will be doubled giving 4. You may take it from there.

An interesting observation to make is that when using the Shell method, backing up is possible during all passes, including the first. Under what conditions is backing up possible during pass 1? How many times may backing up take place during that pass?

EXERCISES

12-1. Discuss the operation of the interchange method of sorting.

12-2. Discuss the operation of the Shell method of sorting.

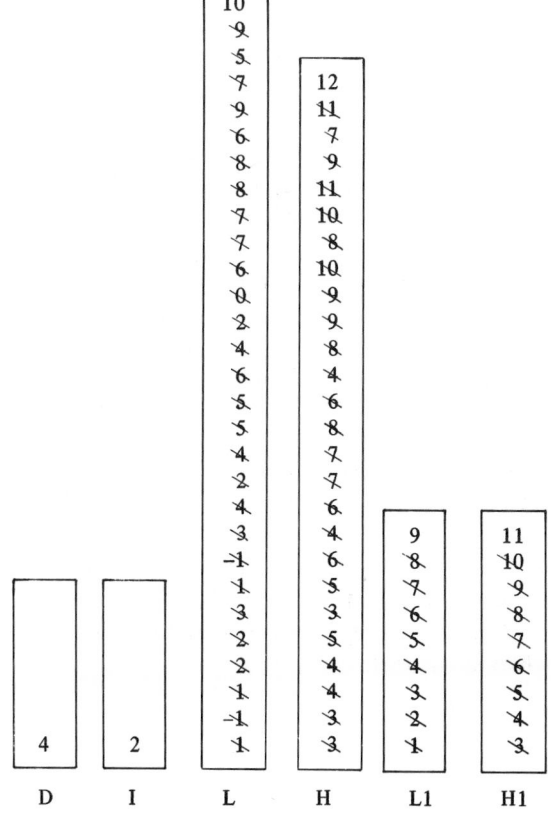

FIG. 12-27

Positions	ACCT-NUMS
1	314
2	425
3	319
4	436
5	441
6	444
7	453
8	467
9	475
10	473
11	484

FIG. 12-28

12-3. Using the interchange method of sorting, how many passes, maximum, are required through a 10,000-cell list in order to sort it in increasing sequence? How many passes are required through the list when using the Shell method?

12-4. What is the purpose of the telltale flag described in the text under the discussion of the interchange method of sorting? Could such a flag also be used when using the Shell method?

12-5. Why is a temporary holding cell needed when swapping the values of two cells?

12-6. Using the Shell sorting method, how many passes, maximum, will be required to sort a list of 1075 values?

12-7. What triggers a "back up" when using the Shell sorting method? How long does backing up continue? Is backing up possible during the first pass?

12-8. Construct a list containing seven values stored in random sequence. Show how the variables D, I, L, H, L1, and H1 change as the list is sorted using the Shell method. The variable names shown here have the same meanings that they have in the text.

Chapter 13

TAPE SORTING

The Shell method of sorting is very efficient if all the values to be sorted can be stored in a computer's memory at the same time. When those values are relatively few, say 200 or 300, or even 2000 or 3000, placing them in memory in a single list is quite feasible. But often there are 50,000 or 60,000 values to be sorted—at times, even millions. Clearly, these values cannot all be placed in memory at the same time; most computers' memories are too small to allow this.

Let us consider an example: suppose that there are 50,000 values recorded on a reel of magnetic tape (individual tapes have the capacity to hold far more values than this). Call the tape TAPE A. Assume that these values are all different (this is guaranteed in order to simplify this discussion) but that they are in no special order. For example, the first few numbers on the tape may be 368947, 26716, 9476387, 6813, 476, 94768, etc. Let us say that the problem is to sort these numbers in increasing sequence and store them on the original reel of tape, TAPE A.

One way that the problem may be solved is to bring into use the memory of a computer and two additional reels of tape. These two tapes, TAPE B and TAPE C; will be used to help sort the values of TAPE A. We shall show soon exactly how these tapes are to be used.

The method to be used will involve bringing into memory several of the values obtained from TAPE A. Some preliminary sorting will be done in memory and sorted sequences of values will be moved to

TAPE B and TAPE C. This procedure will continue until all the values found on TAPE A have been processed. At that time, TAPE B and TAPE C will contain many disjointed sorted sequences. These disjointed sorted sequences are called *strings*.

What is a disjointed sorted sequence (string)? It is a series of values that are in order, either increasing or decreasing as desired. TAPE B might contain, for example, the three strings shown in Fig. 13-1 at the very beginning of the tape. Note that all the values in strings 1, 2, and 3 are in increasing sequence within the strings themselves. The strings are sorted within themselves, but the values in each string are independent of the values in the other strings. We have, in the example, three disjointed sorted strings.

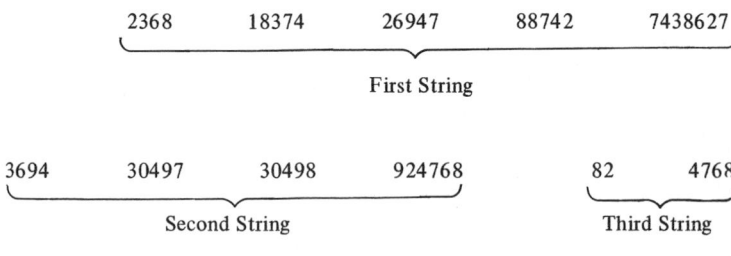

FIG. 13-1

The example shows that strings are of various sizes. The contents of a string may be a single value or hundreds, thousands, or even more values.

The fourth string is not shown in Fig. 13-1, but we do know something about it. The first value of the string must be a value smaller than 4768 (the last value of the third string). How do we know this? Simply because if the next value were larger than 4768, that value would be a member of the third string.

When TAPE A is processed, no attempt is made to equalize the number of strings recorded on TAPE B and TAPE C. In fact, for the sake of efficiency, the numbers of strings placed on those two tapes are deliberately made unequal. You will soon see why this is so.

Once strings have been stored on TAPE B and TAPE C, the next step is to merge the strings on the two tapes and store the merged strings back on TAPE A. Merging means joining in sequence the values from the strings on TAPE B with the values from the strings on TAPE C. To be more precise, the first string of TAPE B is merged with the first string of TAPE C. The merged string is stored on TAPE A. Then the second string of TAPE B is merged with the second string of TAPE C and the merged string becomes the second string stored on TAPE A.

This process continues until the strings on either TAPE B or TAPE C are exhausted. One of the tapes will be exhausted before the other because the numbers of strings on the tapes were deliberately made unequal.

Let us say that the strings on TAPE B run out before those on TAPE C. Now we have strings on TAPE A and TAPE C. The next step is to merge the strings on TAPE A and TAPE C placing merged strings on TAPE B. This procedure continues until the strings on one of the tapes runs out. The strings on TAPE C will be exhausted before those on TAPE A.

You can probably guess the rest of the story. Tapes, TAPE A, TAPE B, and TAPE C, keep juggling the strings and merging them. As strings are merged, they become longer and longer. The end result is a single tape, TAPE A, with a single string containing 50,000 values.

Do not be concerned if all this sounds complicated and confusing. It is complicated but you will understand it clearly when we go through an example step by step.

Before we give the example, we shall explain a concept that you will need to understand in order to follow the solution to the problem.

There is a sequence of numbers known as the Fibonacci series. The series begins this way:

$$0, 1, 1, 2, 3, 5, 8, 13, 21, 34, 55, 89$$

There is no end to the series. As you may have already observed, each number in the series is formed from the sum of the two previous numbers. Thus, 2 is the sum of 1 and 1; 21 is the sum of 8 and 13; 89 is the sum of 34 and 55; etc. The next number in the series shown is 144, this value being the sum of 34 and 55.

Assume that there is a reel of tape, TAPE A, containing 100 numbers in random sequence. These are the numbers (for simplicity, all numbers lie between 1 and 200, inclusive, and all the numbers are different):

187, 83, 177, 161, 158, 82, 186
123, 111, 72, 18, 148, 112, 62
140, 6, 71, 91, 193, 147, 95
116, 180, 129, 179, 198, 1, 3
192, 36, 167, 50, 47, 122, 178
17, 166, 137, 81, 118, 66, 131
170, 27, 154, 150, 5, 86, 9
87, 55, 183, 38, 73, 78, 54
143, 165, 200, 34, 155, 65, 199
185, 144, 136, 195, 125, 23, 32
21, 190, 142, 121, 172, 159, 64
174, 194, 149, 164, 11, 52, 134
7, 25, 102, 135, 184, 90, 151
24, 56, 132, 80, 39, 188, 30
33, 67

We are to sort these numbers in increasing sequence using the Fibonacci tape sort method.

At this point, it is important to make the observation that we have arbitrarily decided to give an example of *only* 100 numbers. This has been done so that the steps involved in sorting the numbers can more easily be followed. In actual situations, the number of numbers on the tape could probably range into the thousands, tens of thousands, even millions. In an actual application, if only 100 numbers were to be sorted, the Fibonacci tape sort method would not be used. Instead, all 100 numbers would be brought into memory, stored in a list, and then sorted using the Shell method or some other efficient method. The Fibonacci tape sort method is used when all numbers stored on the source (in this example, a reel of tape) cannot all be brought into memory at one time.

The first step in sorting the 100 numbers is to create strings of sorted numbers. This is accomplished by using a string-generating scheme called the tournament method. Here is how the tournament method works:

First, 16 numbers are brought in from the source and stored in 16 cells of memory. The 16 numbers from the source as given in this example are given in Fig. 13-2.

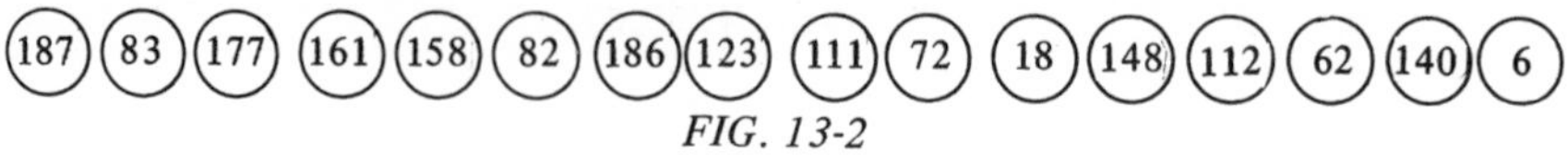

(187)(83)(177)(161)(158)(82)(186)(123)(111)(72)(18)(148)(112)(62)(140)(6)

FIG. 13-2

The next step is to compare the numbers in pairs to find *winners* of all pairs. Thus, 187 is compared with 83; 177 is compared with 161; etc. After eight compares have been made, the winners are stored in eight cells of memory, as shown in Fig. 13-3. The winner of each pair is the smaller of the two values in the pair.

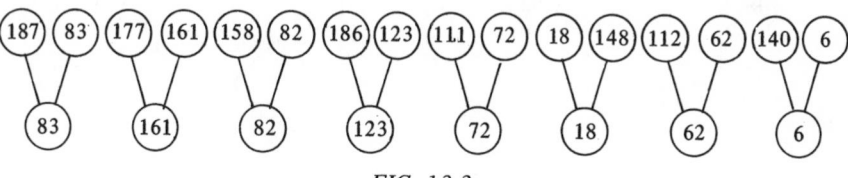

FIG. 13-3

The next step is to compare the values in the shorter series, pair by pair, and establish a list of four winners, as in Fig. 13-4.

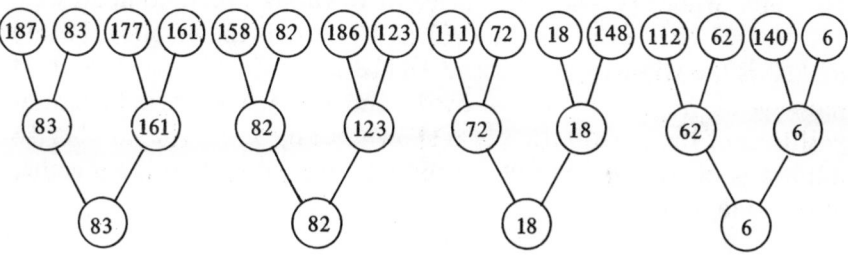

FIG. 13-4

You can see the pattern that is emerging. The next step is to compare 83 against 82 to find a winner and 18 against 6 to find a winner. Then the one winner, 82, is compared against the other, 6, to establish the final winner through the tournament, 6.

The complete chart showing that 6 is found to be the smallest of the first 16 numbers examined is given in Fig. 13-5. The number, 6, becomes the first number of a *string*. This number is written on TAPE B. Now we shall follow 6 with another number. If that number is larger than 6, it belongs to the first string. If it is less than 6, it begins the second string.

When a number in the list of 16 numbers has been found to be smallest, that number is replaced by another from the source and the tournament is repeated. The next number in the source is 71. That number takes the place of the 6 in the list of 16 numbers. The

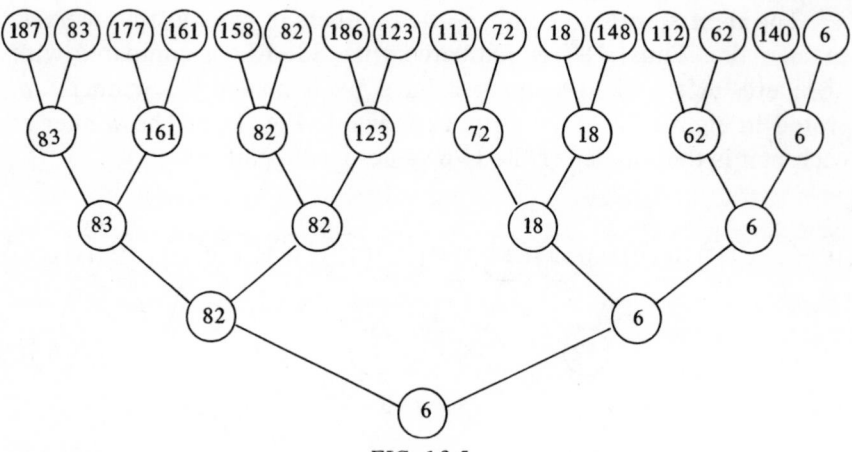

FIG. 13-5

tournament resumes, but only required comparisons are made. There is no need to compare 187 with 83 again, for example, nor 177 with 161, etc. When 71 takes the place of 6, the required compares are 140 and 71 (71 is the winner); 62 and 71 (62 is the winner); 18 and 62 (18 is the winner); and 82 and 18 (18 is the final winner). A chart showing the replacement of 6 by 71 and the new winners in various positions of the organization is as shown in Fig. 13-6. The number 18 follows 6 stored on TAPE B. Since 18 is larger than 6, it is a member of the same string.

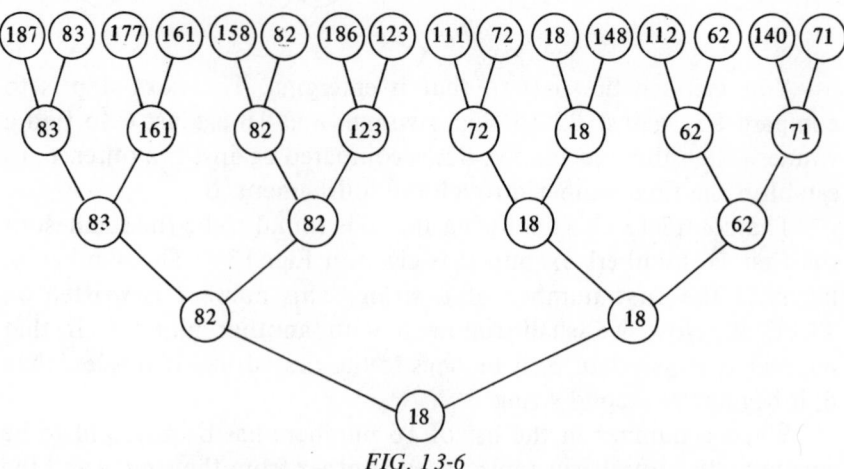

FIG. 13-6

Let us go through the tournament once more. The number from the source that takes the place of 18 is 91. Again, only the necessary compares of the tournament are performed. These compares involve 91 against 148 (91 is the winner); 72 against 91 (72 is the winner); 72 against 62 (62 is the winner); and 82 against 62 (62 is the final winner).

A chart showing the replacement of 18 by 91 and the new winners in various positions of the organization is shown in Fig. 13-7. The number 62, being larger than 18, becomes a member of the first string. It is added to TAPE B. Whenever a number stored on TAPE B is smaller than the last number stored there, it becomes the first number of a new string.

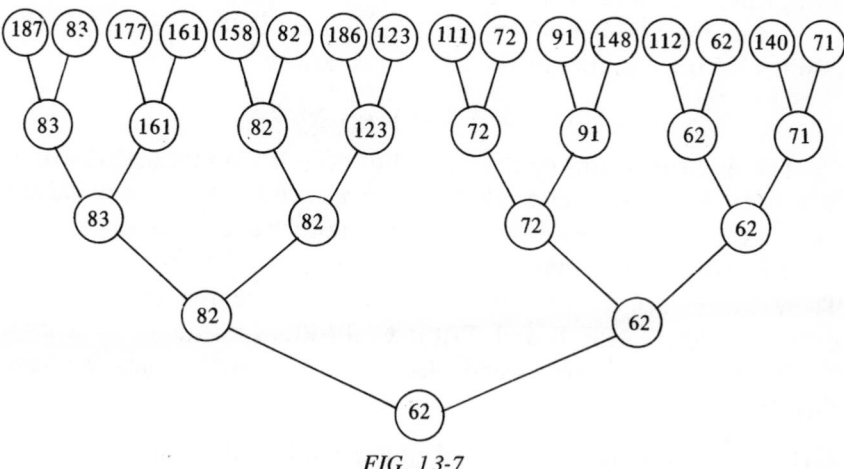

FIG. 13-7

The process given above continues until all values from the source have been processed and have become members of strings. Strings are stored on TAPE B as they are created. When all strings have been generated, TAPE B looks like this:

```
6 18 62 71 72 82 83 91 95 111 112/ 1 3
116/ 36 123/ 50/ 47 122 129/ 17 140/ 137/
81 118/ 66 131 147/ 27 148 150/ 5 86/
9 87/ 55 154/ 38 73 78/ 54 143 158 161/
34 155/ 65 165 166/ 144/ 136 167/ 125/ 23 32/
21 170/ 142/ 121 172/ 159/ 64 174 177/
149 164/ 11 52 134/ 7 25 102 135 178/
90 151/ 24 56 132/ 80/ 39 179/ 30 33 67
180 183 184 185 186 187 188 190 192 193 194
195 198 199 200/
```

We have placed slashes in the list of numbers on TAPE B to show where one string leaves off and another begins.

Examine the strings. You will see that a string may contain as few as one number. It may, of course, contain any number of numbers. The number of numbers in a string partially depends on how scrambled the numbers were on the original source, TAPE A. The sizes of the strings also depend on how many values are used in the original tournament list. We used only 16, but any power of 2 is acceptable: 32, 64, 128, 256, 512, 1024, etc. The more computer memory is available for the tournament list, the longer the resulting strings and the more efficient the sort.

Count the number of strings. You will see that there are 34. The number 34 is a member of the Fibonacci series:

$$0, 1, 1, 2, 3, 5, 8, 13, 21, 34, 55, 89$$

This is a fortuitous coincidence, but even if the number of strings had been something else, *dummy* strings could have been added forcing the total to be a value found in the Fibonacci series.

The two values that, when summed, equal 34 are 13 and 21. These two values tell how the 34 strings must be divided and stored on two tapes, TAPE B and TAPE C. TAPE B contains 34 strings. Transfer 21 of the strings and place them on TAPE C and we have this situation:

TAPE A   Contains the original random values (we shall call this a "blank" tape for reasons that will become apparent later).

TAPE B   Contains 13 strings.

TAPE C   Contains 21 strings.

Here are the strings found on TAPE B and TAPE C:

TAPE B

         6 18 62 71 72 82 83 91 95 111 112/
         1 3 116/
         36 123/
         50/

47 122 129/
17 140/
137/
81 118/
66 131 147/
27 148 150/
5 86/
9 87/
55 154/

TAPE C

38 73 78/
54 143 158 161/
34 155/
65 165 166/
144/
136 167/
125/
23 32/
21 170/
142/
121 172/
159/
64 174 177/
149 164/
11 52 134/
7 25 102 135 178/
90 151/
24 56 132/
80/
39 179/
30 33 67 180 183 184 185 186 187 188 190 192 193 194
195 198 199 200/

On TAPE B and TAPE C the strings follow each other without breaks. However, so that we may more easily discuss the steps involved in the sort, we are showing the strings beginning on separate lines.

The next step is to *merge* the strings found on TAPE B and TAPE C and place the merged strings on TAPE A. Let us look at the first string of TAPE B and the first string of TAPE C.

TAPE B

6 18 62 71 72 82 83 91 95 111 112/

TAPE C

38 73 78/

Since both strings are in sequence within themselves, all members of both strings *do not* have to be in memory in order to sequence them. Small portions of the first and second strings are examined. As the new merged string is formed, it is written on TAPE A.

TAPE A gets the merged form of the two strings. That merged string is

TAPE A

6 18 38 62 71 72 73 78 82 83 91 95 111 112/

Observe that the total number of members in this string is 14. (There were 11 members in one string and 3 members in the other.)

Now, the second string of TAPE B and TAPE C are merged. Those strings are

TAPE B

1 3 116/

TAPE C

54 143 158 161/

TAPE A gets

1 3 54 116 143 158 161/

How long does this process continue? Until all the strings available on TAPE B have been processed. At that point, we shall say that TAPE B is *blank*. It really is not, but we may consider it so. When TAPE B has been made blank, TAPE C still contains 8 unprocessed strings (21 there originally, less 13 merged with 13 strings from TAPE B). TAPE A contains 13 merged strings.

Here is a summary of the situation that exists when TAPE B has been made "blank":

TAPE A   Contains 13 merged strings (from TAPE B and TAPE C). The information originally recorded on TAPE A was erased when new material was written on it.

TAPE B   Is blank.

TAPE C   Contains 8 strings that (as yet) did not get a chance to be merged with any other strings.

Here are the strings found on TAPE A and TAPE C.

TAPE A

6 18 38 62 71 72 73 78 82 83 91 95 111 112/
1 3 54 116 143 158 161/
34 36 123 155/
50 65 165 166/
47 122 129 144/
17 136 140 167/
125 137/
23 32 81 118/
21 66 131 147 170/

27 142 148 150/
5 86 121 172/
9 87 159/
55 64 154 174 177/

TAPE C

149 164/
11 52 134/
7 25 102 135 178/
90 151/
24 56 132/
80/
39 179/
30 33 67 180 183 184 185 186 187 188 190 192 193
        194 195 198 199 200/

TAPE A has 13 strings; TAPE C has 8. These are two consecutive numbers found in the Fibonacci series:

$$0, 1, 1, 2, 3, 5, 8, 13, 21, 34, 55, 89$$

This is no coincidence. Once the original number of strings (in this example, 34) have been divided according to the Fibonacci sequence, tapes will contain numbers of strings found in the Fibonacci sequence as strings are merged. (In general, as sorting progresses, tapes will contain fewer and fewer strings but those strings will become longer and longer.)

The next step is to merge the strings found on TAPE A and TAPE C on to the "blank" tape, TAPE B.

Merging will be performed until the strings on TAPE C are exhausted. At this time we shall define TAPE C to be "blank." TAPE B will then contain 8 strings (merged from TAPE A and TAPE C) and TAPE A will contain 5 strings (which did not, as yet, get a chance to be merged with any other strings).

Predictably, the nonblank tapes, TAPE A and TAPE B, contain 5 and 8 strings, respectively, two numbers found in the Fibonacci series. The contents of TAPE A and TAPE B are these:

TAPE A

21 66 131 147 170/
27 142 148 150/
5 86 121 172/
9 87 159/
55 64 154 174 177/

TAPE B

6 18 38 62 71 72 73 78 82 83 91 95 111 112 149 164/
1 3 11 52 54 116 134 143 158 161/
7 25 34 36 102 123 135 155 178/
50 65 90 151 165 166/
24 47 56 122 129 132 144/
17 80 136 140 167/
39 125 137 179/
23 30 32 33 67 81 118 180 183 184 185 186 187 188 190
192 193 194 195 198 199 200/

We shall not show the details of further mergings, but you would find it educational if you worked them out for yourself. A summary showing how the cascading of strings from tape to tape takes place until all the strings are sorted is given in Figs. 13-8 through 13-15. The tapes take turns being "blank," TAPE A, TAPE B, TAPE C, TAPE A, etc. The number of strings on the tapes become fewer and fewer until two tapes have only one string each. The final phase merges the two strings and the sorting is completed. This is the final string found on TAPE A:

1 3 5 6 7 9 11 17 18 21 23 24 25 27 30 32
33 34 36 38 39 47 50 52 54 55 56 62 64
65 66 67 71 72 73 78 80 81 82 83 86 87
90 91 95 102 111 112 116 118 121 122 123 125
129 131 132 134 135 136 137 140 142 143 144 147 148 149
150 151 154 155 158 159 161 164 165 166 167 170 172 174
177 178 179 180 183 184 185 186 187 188 190 192 193 194
195 198 199 200

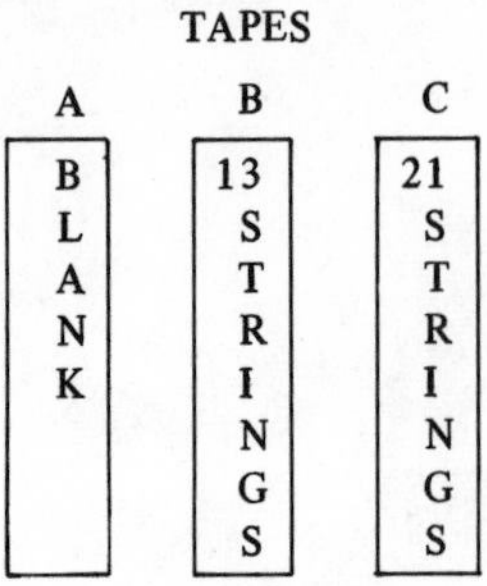

TAPES

A	B	C
B L A N K	13 S T R I N G S	21 S T R I N G S

FIG. 13-8

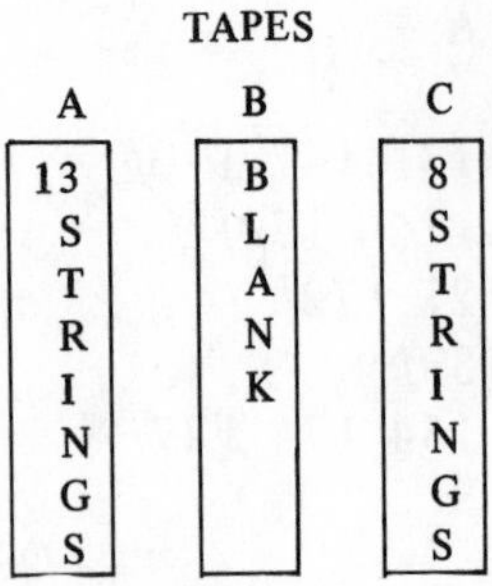

TAPES

A	B	C
13 S T R I N G S	B L A N K	8 S T R I N G S

FIG. 13-9

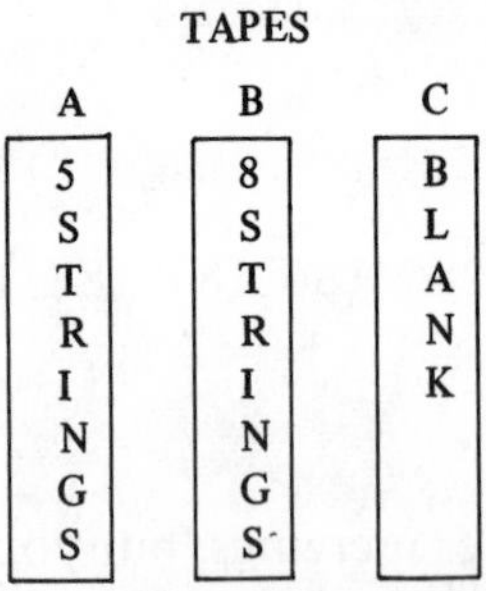

TAPES

A	B	C
5 S T R I N G S	8 S T R I N G S·	B L A N K

FIG. 13-10

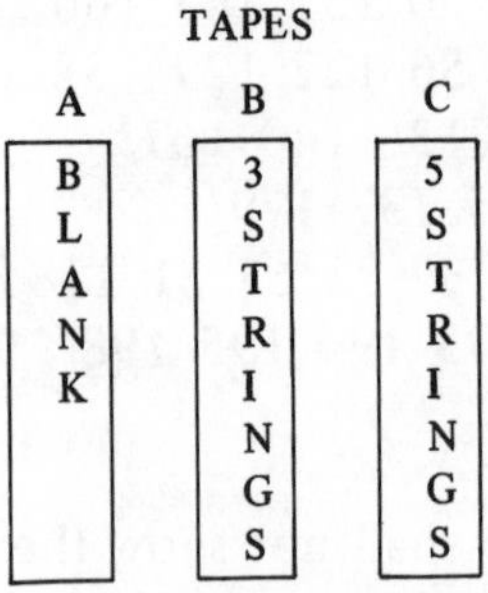

TAPES

A	B	C
B L A N K	3 S T R I N G S	5 S T R I N G S

FIG. 13-11

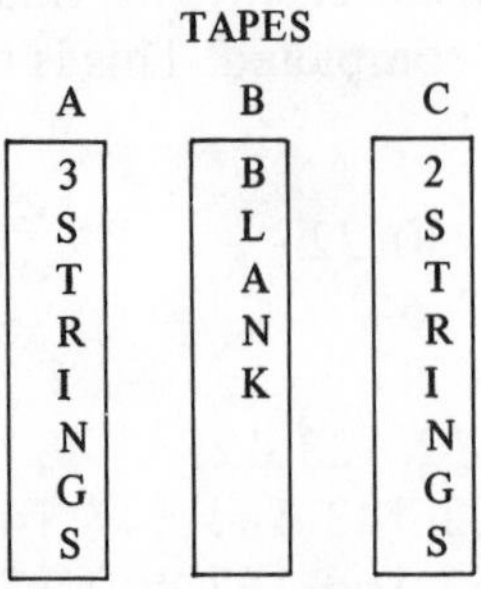

TAPES

A	B	C
3 S T R I N G S	B L A N K	2 S T R I N G S

FIG. 13-12

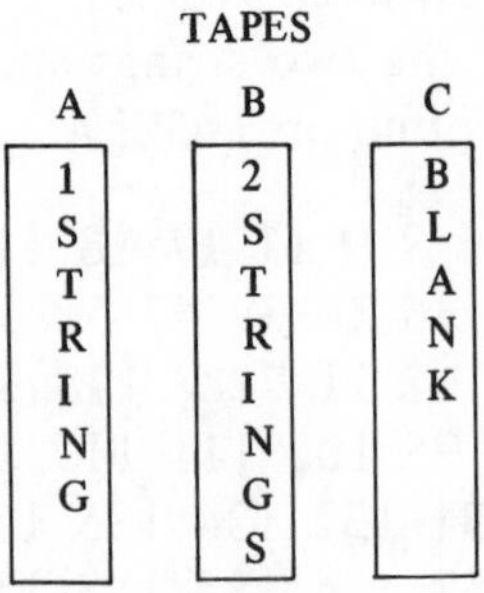

TAPES

A	B	C
1 S T R I N G	2 S T R I N G S	B L A N K

FIG. 13-13

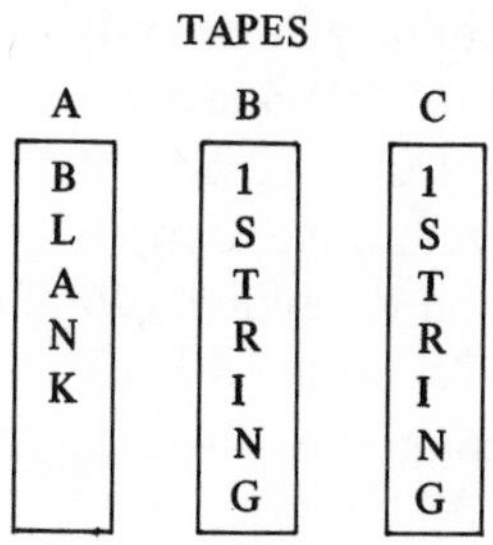

TAPES

A	B	C
B	1	1
L	S	S
A	T	T
N	R	R
K	I	I
	N	N
	G	G

FIG. 13-14

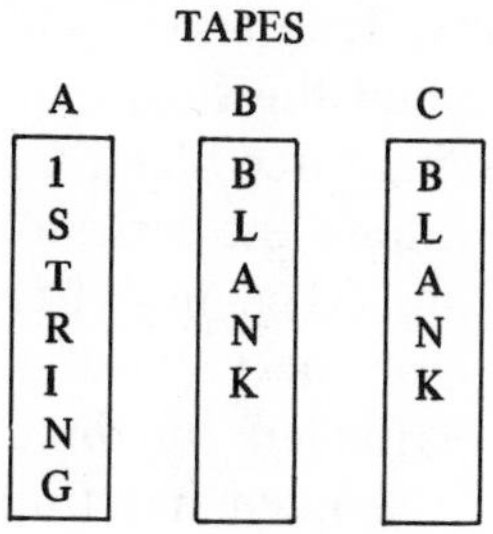

TAPES

A	B	C
1	B	B
S	L	L
T	A	A
R	N	N
I	K	K
N		
G		

FIG. 13-15

If you will count the number of string pairs that were merged, you will find that 33 pairs were merged, one less pair than the total number of original strings. This relationship will always be true. When there are 89 original strings, 88 pairs of strings will be merged; when there are 144 strings, 143 pairs of strings will be merged; etc.

The Fibonacci tape sort method becomes more and more efficient as more and more tapes are assigned to process the strings. Suppose, for example, that four tapes are available instead of three. We shall call the tapes TAPE A, TAPE B, TAPE C, and TAPE D.

Now let us assume that the number of strings that have been generated from the tournament and that must be allocated among three tapes (leaving one blank) is 57. The correct distribution of these strings calls for TAPE B to receive 13 strings, TAPE C to receive 20 strings, and TAPE D to receive 24 strings. The series from which these quantitites are obtained is this:

```
1 1 1
  1 2 2
    2 3 4
      4 6 7
        7 11 13
          13 20 24
            24 37 44
              44 68 81
                etc.
```

The initial distribution of strings is 13, 20, and 24 (57 strings). After merging has taken place, the tapes will contain 7, 11, and 13 strings; the next time, 4, 6, and 7; the next time, 2, 3, and 4; the next time, 1, 2, and 2; etc. Eventually, there will be only one string per tape and the final merging will result in a single string on a single tape leaving three blank tapes. The sort will then be completed.

To extend the series of numbers shown above, copy 81 and then add 44 and 81 giving 125. Next add 68 and 81 giving 149. The three numbers of the extension will be 81, 125, and 149. (The series is always extended by *three* numbers.) Using this pattern, the series may be extended indefinitely.

Where four tapes are used in the sort, the strings are not merged as pairs but as triplets. Thus, if TAPE B, TAPE C, and TAPE D have to be merged and stored on TAPE A, the first string of TAPE B will be merged with the first string of TAPE C and the first string of TAPE D. The result will be stored as the first string of TAPE A. The process will then be continued until one of the tapes runs out of strings.

Suppose, for example, that the first strings of TAPE B, TAPE C, and TAPE D were these:

TAPE B

6 14 18 32/

TAPE C

3 4 50/

TAPE D

13 21 22 33 35/

The first string stored on TAPE A would be

3 4 6 13 14 18 21 22 32 33 35 50/

It would be an interesting exercise for you to determine what series of numbers are involved when five, six, seven tapes, etc., are used in sorting. Here, for example, is the beginning of the series to be used when five tapes are available:

```
1  1  1  1
   1  2  2  2
      2  3  4  4
         4  6  7  8
               etc.
```

and here is the beginning of the series when six tapes are used:

```
1  1  1  1  1
   1  2  2  2  2
      2  3  4  4  4
         4  6  7  8  8
                  etc.
```

See if you can deduce the rule required to generate the numbers in the series.

EXERCISES

13-1. Why cannot the Shell sorting method always be used?

13-2. Define the term *string*.

13-3. Write the first 20 terms of the Fibonacci number series.

13-4. What is the function of the tournament in the Fibonacci tape sort method?

13-5. What is the fewest number of tapes that may be employed when the Fibonacci tape sort method is used? Would the sort be more efficient if more tapes could be employed?

13-6. What is the Fibonacci-like sequence of numbers to use in the Fibonacci tape sort method when seven tapes are used?

ANSWERS
TO ODD-NUMBERED
EXERCISES

CHAPTER 1

1-1. (a) $W = 4/5$ (b) $X = 2$
(c) $Z = 5$ (d) $R = 5$
(e) $T = 1$ (f) $Z = -352/315$
(g) $x = 1479/308$
(h) $R = 40/11$ (i) $X = 15$

(j) $X = \dfrac{-7 \pm \sqrt{721}}{28}$

1-3. (a) $x = 0, y = 4$
(b) $x = 16, y = -12$
(c) $x = 26, y = 20$
(d) $x = -49/22, y = -25/22$
(e) $x = 9/13, y = -19/13$
(f) indeterminate
(g) $x = 16, Y = 6$
(h) $x = 17/9, y = -4/9$
(i) $R = 7/3, S = -5/6$
(j) $X = 0, Y = 2C/b$

1-5. (a) 30   (b) 1536
(c) 25   (d) 512
(e) 78
(f) $2^{30} = 1,073,741,824$
(g) $1/256$ (h) -39
(i) 512   (j) -1

1-7. The data for the first 7 months forms an arithmetic progression. Hence the forecast for the last 5 months might be $31, 34, 37, 40,$ and 43 units respectively. Total sales for the year would be 318.

1-9. \$1,176,804.36 or $10(1.03)^{395}$

CHAPTER 2

2-1. (a) $324_6 = 124_{10}$ (c) $3_6 = 3_{10}$
(e) $11001_6 = 1513_{10}$
(g) $30003_6 = 3891_{10}$
(i) $12345_6 = 1865_{10}$

2-2. (a) $100_5 = $ 25_{10}
(c) $3_5 = $ 3_{10}
(d) $444_5 = $ 124_{10}
(g) $3410032_5 = 60017_{10}$

2-3. (a) $8_{16} = \quad 8_{10}$
     (c) $ABC_{16} = 2748_{10}$
     (e) $8000_{16} = 32768_{10}$
     (g) $1010_{16} = \quad 4112_{10}$

2-4. (a) $99_{10} = \quad 143_8$
     (c) $1001_{10} = 1751_8$
     (e) $9_{10} = \quad 11_8$
     (g) $8_{10} = \quad 10_8$

2-5. (a) $16_{10} = \quad 10_6$
     (c) $24_{10} = \quad 18_{16}$
     (e) $99_{10} = \quad 63_{16}$
     (g) $1101_{10} \quad 44D_{16}$

2-6. (a) $3504_6 = 832_{10}$
     (c) $3B6_{16} = 950_{10}$
     (e) $1011_2 = \quad 11_{10}$
     (g) $1246_7 = 475_{10}$
     (i) $684_{12} = 964_{10}$

2-7. (a) $33_6 = \quad 41_5$
     (c) $AB6_{16} = 222312_4$
     (e) $3777_8 = \quad 2047_{10}$
     (g) $824_{20} = \quad CAC_{16}$
     (i) $11111_2 = \quad 1F_{16}$

2-8. (a) $.34_{10} = .01010111_2^{+}$
     (c) $.05_{10} = .0CCC_{16}^{+}$

2-9. In the number 8478_{10}, there
    are
        8 one-thousands
        4 one-hundreds
        7 tens
        8 units

2-11. (a) $3452_6 = \quad 824_{10}$
     (c) $500040_6 = 38904_{10}$

2-13.

etc. $\Big\{$ | 6561 | 729 | 81 | 9 | 1 |

2-15. (a) $3074_{10} = \quad 6002_8$
     (c) $6EC4_{16} = 28356_{10} =$
        145446_7

2-17.

	0	1	2	3	4
0	0	1	2	3	4
1	1	2	3	4	10
2	2	3	4	10	11
3	3	4	10	11	12
4	4	10	11	12	13

2-19. $1234_5 + 3004_5 + 2103_5 +$
      $0323_5 + 4112_5 = 21341_5$

2-21. (a) 53_{10}
     (c) 29_{10}

2-23. Product is 00100110100
      (308_{10})

2-25. 9.15625_{10}

2-27. 1010001.10001_2

CHAPTER 3

3-1. (a) 00000000000000000000-
        111011000001_2
     (c) 00000000000000000000-
        000001000000_2
     (e) 00000000000000000000-
        000001100011_2

3-2. (a) 11111111111111111111-
        111111100000_2
     (c) 11111111111111111111-
        111110000101_2
     (e) 11111111111111111111-
        111111111101_2

3-3. (a) 00011110_2 (30_{10})
(c) 00001110_2 (14_{10})
(e) 11110010_2 (-14_{10})

3-5. A K equals 1024 words.

3-7. A byte is a group of 8 bits.

3-9. 66609_{10}

3-11. 05664057006_8

3-13. 000000000001100000110-
101001111101_2

3-15. 1000 (-8_{10})
1001 (-7_{10})
1010 (-6_{10})
1011 (-5_{10})
1100 (-4_{10})
1101 (-3_{10})

1110 (-2_{10})
1111 (-1_{10})
0000 (0_{10})
0001 (1_{10})
0010 (2_{10})
0011 (3_{10})
0100 (4_{10})
0101 (5_{10})
0110 (6_{10})
0111 (7_{10})

3-17. (-8_{10})

3-19. Characteristic and mantissa

3-21. 00000101_2

3-23. $C3_{16}$ and FA_{16}

3-25. 400000_{16}

CHAPTER 4

4-1. $P = D7_{16} = 11010111_2$
$S = E2_{16} = 11100010_2$
$4 = F4_{16} = 11110100_2$
$\$ = 5B_{16} = 01011011_2$

4-3. READ

4-5. READ DATA CARD AND PLACE CONTENTS BEGINNING AT BYTE LOCATION 1000.

CONVERT BYTES 1000-1004 TO INTEGER AND STORE RESULT AT WORD LOCATION 2000.

CONVERT BYTES 1005-1009 TO INTEGER AND STORE RESULT AT WORD LOCATION 2004.

MULTIPLY INTEGER 2000 BY 2004 GIVING 2008.

CONVERT INTEGER 2008 TO BYTES AND STORE RESULT AT 4000-4005.

PRINT 4000-4005 BEGINNING AT PRINT POSITION 20 SUPPRESSING LEADING ZEROES.

STOP RUN.

4-7. Compilation is the name of the process during which programs written in high-level languages such as PL/I, COBOL and FORTRAN are converted to computer language.

CHAPTER 5

5-1. List, table

5-3. A subscript is an integer (whole number) value which points to some specific location of an array.

5-5. There is no limit.

CHAPTER 6

6-1. (a) $A = \{x; x \in P, 3 < x < 8\}$
(b) $B = \{x; x \in P, 19 < x < 25\}$
(c) $C = \{11,12,13,14,15,16,17, 18,19\}$
(d) $D = \{1001\}$

6-3. (a) $A \cap B = \{x\}$
(b) $A \cap C = \{z\}$
(c) $B \cap C = \{b\}$
(d) $A \cup B = \{a,b,x,y,z\}$
(e) $A \cup C = \{b,x,y,z\}$
(f) $B \cup C = \{a,b,x,z\}$

6-5. (a) $A' = \{x; x \in P, x < 5$ or $x > 8\}$
(b) $B' = \{1\}$
(c) $C' = P$
(d) $D' = \{1,2,3,4,5\} = \{x; x \in P, x < 6\}$
(e) $E' = \{x; x \in P, x$ not divisible by 5$\}$

6-7. (a) $A \cup C = \{x; x \in P, x < 3$ or $x > 5\}$
(b) $A \cup D = \{x; x \in P, x < 9\}$
(c) $B \cup C = C$
(d) $B \cup D = \{x; x \in P, x < 9\}$
(e) $C \cup D = P$
(f) $A \cup B \cup C = \{x; x \in P, x < 3$ or $x > 4\}$
(g) $A \cup B \cup C \cup D = P$

6-9. (a)

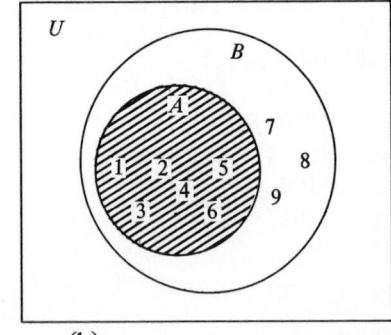

(b)

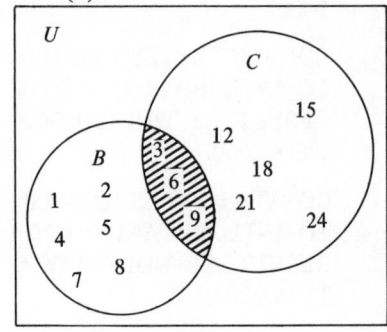

(c)

(d)

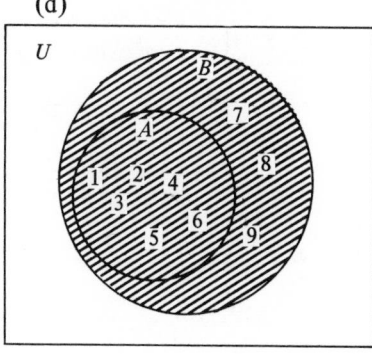

(f)

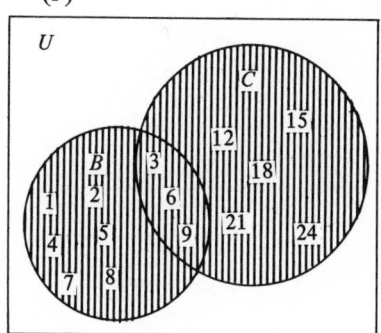

(e)

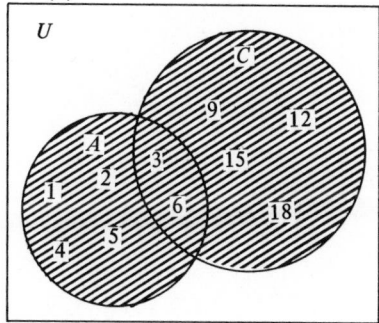

(g) See (a)

6-11. (a) $(A \cup B) \cap (C \cup D) = \{1,2,3,4,5\}$
(b) $(A \cap B) \cup (C \cap D) = \{\emptyset\}$
(c) $(A \cup B \cup C) \cap D = \{1,2\} = D$
(d) $(A \cap B \cap C) \cup D = \{1,2\} = D$
(e) $(A \cap B) \cup (A \cap C) \cup (A \cap D) = \{1,2,3,4\} = A$
(f) $(A \cup B) \cap (A \cup C) \cap (A \cup D) = \{1,2,3,4\} = A$

CHAPTER 7

7-1.

IF				
AGE IS 20	Y	Y	N	N
FRESHMAN	Y	N	Y	N
THEN				
INCREMENT TOTAL RECORD COUNTER	Y	Y	Y	Y
INCREMENT "HIT" COUNTER	Y	N	N	N
PROCESS RECORD	Y	N	N	N
PRINT RECORD	Y	N	N	N

7-3. (a)

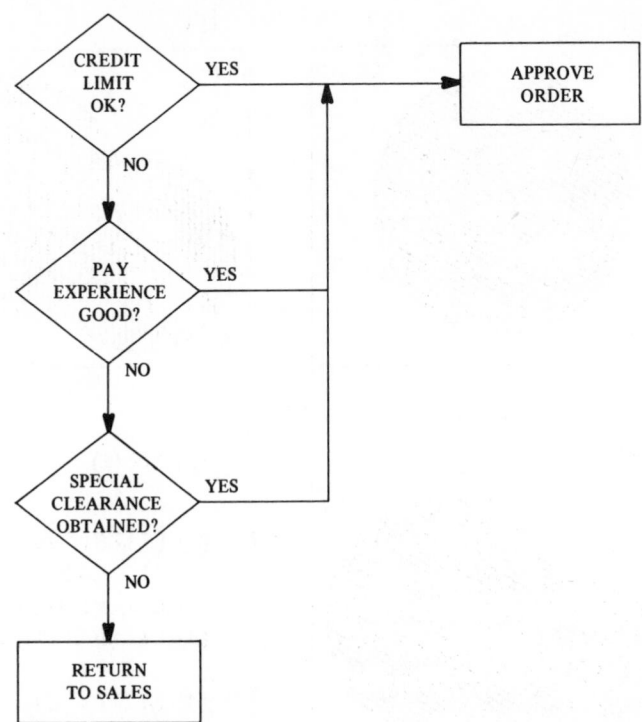

(b)

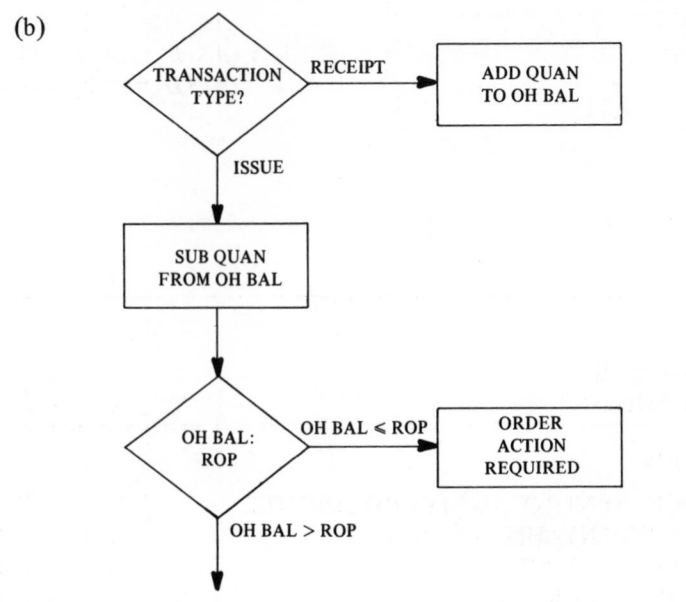

CHAPTER 8

8-1. 2 reels

8-3.

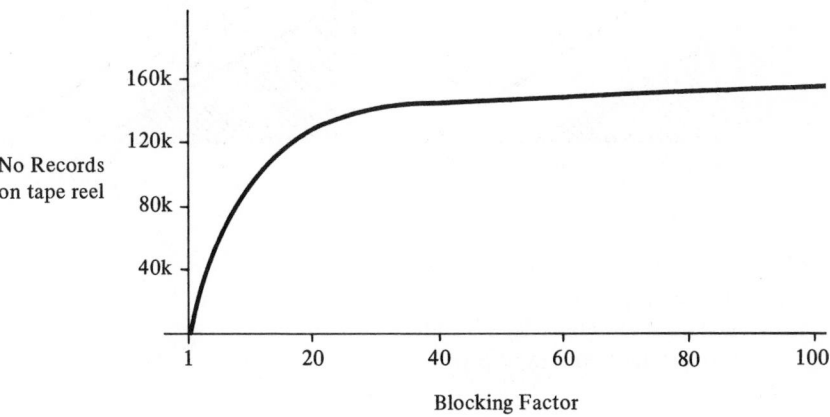

No Records on tape reel

Blocking Factor

CHAPTER 9

9-1. (a) $x > 7/2$ (b) $x \leqslant 7/2$
(c) $x > 3/2$ (d) $-1 < x$
(e) $x > 1$ and $x < 3$
(f) $x < -7$
(g) No Solution
(h) $x \geqslant -5$ and $x \leqslant 5$
(i) $y < 3/2$ (j) $y < -25/4$
(k) $y > 27/50$

9-3. (a)

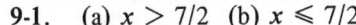

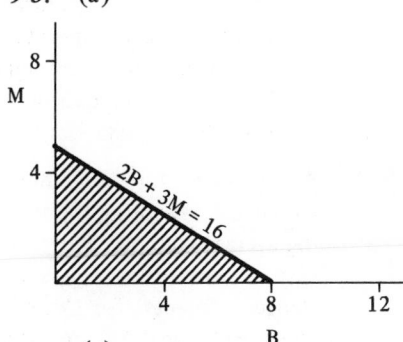

(b)

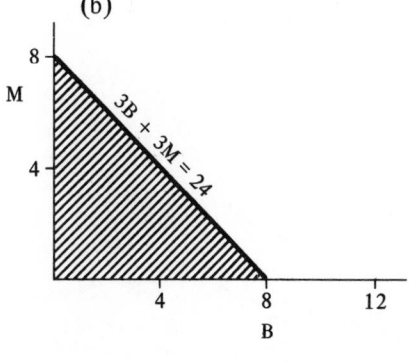

(c)

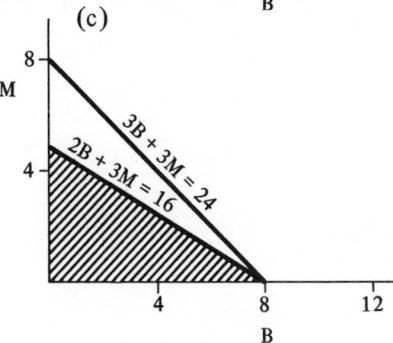

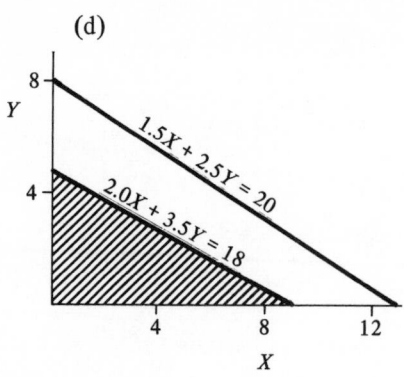

(d)

(e)

9-5. 44P and 16Q produces a maximum profit of \$380 each month. Note also that 0P and 38Q produces the same profit.

CHAPTER 10

10-1. (a) MA = 28.
(b) WMA = 35 using weights of 5, 5, 20, 30, and 40, however the moving average would be preferable because no trend is indicated by the data.

10-3. A forecast for September would be 26.4 using the moving average method. This forecast is within the tolerance. Hence a forecast for October could be calculated by this method.

10-5. Using the second-degree equation
$$y = 5x^2 - 30x + 65,$$
we see that
$$y_{May} = 40.$$
Using the moving average method we see that
$$y_{May} = 25,$$
which clearly better approximates the actual data.

10-7. $y = 3x + 18$

CHAPTER 11

11-1. Knowing the position that a value has in a list permits the accessing of values in one or more related lists.

11-3. A diamond-shaped symbol indicates a decision point.

11-5. Input data

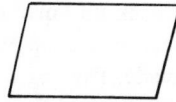

Process data

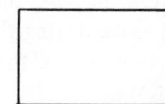

Print data

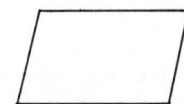

11-7. The subscript is placed within parentheses following the name of the list.

11-9. 1. The search is conducted to the very end of the list and the sought-for value is not found.

2. At any point during the search of a list sorted in increasing sequence, the value being sought is found to be larger than the last list value examined.

11-11. 18

11-13. Assume list values are:

Pos.	
1	3
2	8
3	11
4	15
5	19
6	21
7	27
8	28
9	35
10	37
11	51
12	52
13	53
14	59
15	64
16	67
17	69
18	85
19	87
20	90
21	94
22	95
23	97
24	98
25	99

Assume the value of X is 30.

L	H
1	25
7	12
8	8
9	

SUBSCR

13
6
9
7
8

11-15. So that the binary search program may determine whether to continue a search in the list ahead of the last unsuccessful point looked at or below that point.

CHAPTER 12

12-1. The interchange method requires, as a maximum, N-1 passes through a list of values to be sorted, where N represents the number of values in the list. Each pass through a list is shorter than the one preceding it. The interchange method is not as efficient as the SHELL method.

12-3. 9,999; 13

12-5. To prevent one of the two values being destroyed. That is, when the value at X(N+1) is moved to X(N), the value of X(N) is destroyed unless it has been previously saved.

12-7. A "back up" may be triggered whenever a swap of two values is made. Backing up continues as long as additional swaps are caused by any backing up motion. Backing up is possible only once during the first pass and then only if the size of the list is odd.

CHAPTER 13

13-1. The SHELL sorting method may be used only if all the values to be sorted may be stored in the memory of the computer at one time.

13-3. 0 1 1 2 3 5 8 13 21
       34 55 89 144 233 377
       610 987 1597 2584 4181

13-5. The fewest number of tapes is 3. The more tapes that may be used, the more efficient the sort.

INDEX